Lecture Notes in Mathematics

Edited by A. Dold and B. E

T0255301

1058

Bernd Aulbach

Continuous and Discrete Dynamics near Manifolds of Equilibria

Springer-Verlag
Berlin Heidelberg New York Tokyo 1984

Author

Bernd Aulbach
Mathematisches Institut der Universität
Am Hubland, 8700 Würzburg, FRG

Sponsored by the Volkswagenwerk Foundation.

AMS Subject Classification (1980): 34-02, 34C05, 34C20, 34C30, 34C35, 34C45, 34D05, 34D20, 39A10, 39A11, 58F10, 58F15, 58F19, 92A10

ISBN 3-540-13329-1 Springer-Verlag Berlin Heidelberg New York Tokyo
ISBN 0-387-13329-1 Springer-Verlag New York Heidelberg Berlin Tokyo

Printing and binding: Beltz Offsetdruck, Hemsbach/Bergstr.
2146/3140-543210

Dedicated to the memory

of the first mathematician I met,

my father

Preface

In the qualitative theory of dynamical systems nonisolated equilibria are rarely studied objects in comparison with isolated equilibria. This research monograph is devoted to the analysis of solutions of nonautonomous ordinary differential equations or autonomous ordinary difference equations near manifolds of stationary solutions. This study is not only motivated from a theoretical but also from a practical point of view as an application to the fundamental selection model of population genetics shows.

The principal results of this book can be comprehended with a basic knowledge of ordinary differential equations. A full understanding and appreciation of this text, however, requires some familiarity with several concepts of the qualitative theory of dynamical systems such as ω - limit sets, invariance, Ljapunov functions, stability, asymptotic behavior and invariant manifolds. Apart from this, the presentation is self - contained. All quoted results are stated in full detail.

I am most grateful to Prof. Dr. H. W. Knobloch for his helpful criticism and suggestions for improvement. Many thanks are also due to Dr. D. Flockerzi for his careful reading of the manuscript. Anyhow, all remaining faults are mine. Finally I would like to thank the Volkswagenwerk foundation for financial support during the preparation of the final version of this book.

Würzburg, February 1984 Bernd Aulbach

Table of Contents

Basic Notation

$\mathbb{R}^n$	n-dimensional real space,
$\mathbb{R}^n_+$	nonnegative orthant $\{(x_1,\ldots,x_n) \in \mathbb{R}^n : x_i \geq 0,\ i=1,\ldots,n\}$,
$\mathbb{R}^{m\times n}$	set of real m×n matrices,
$\mathbb{R}^+$	set of nonnegative real numbers,
$\mathbb{N}_o$	set of nonnegative integers,
$C(A,B)$	set of continuous functions from A into B,
$C^k(A,B)$	set of functions from A into B which have continuous derivatives up to order k,
$f_x(x_o)$	Jacobian matrix of $f(x)$ at x_o,
$\dot{x}$	derivative of x with respect to t,
$\times$	Cartesian product or multiplication sign,
M^T	transpose of the matrix M,
e	vector $(1,\ldots,1)^T$ of appropriate dimension,
$x > y$	for vectors $x=(x_1,\ldots,x_n)$, $y=(y_1,\ldots,y_n)$: $x_i > y_i$, $i=1,\ldots,n$,
$x \geq y$	for vectors $x=(x_1,\ldots,x_n)$, $y=(y_1,\ldots,y_n)$: $x_i \geq y_i$, $i=1,\ldots,n$,
$M \geq N$	for square matrices: $M-N$ positive semidefinite,
I_n	n×n identity matrix,
$O_{m\times n}$	m×n zero matrix,
O_m	m×m zero matrix,
$\Vert\cdot\Vert$	arbitrary vector norm and corresponding operator norm,
$\Vert\cdot\Vert_2$	Euclidean norm,
$\Vert\cdot\Vert_*$	matrix norm $\Vert M\Vert_* = (\text{trace } M^TM)^{1/2}$,
B_ρ	open ball $\{x \in \mathbb{R}^n : \Vert x\Vert_2 < \rho\}$ in some $\mathbb{R}^n$,
$\emptyset$	empty set,
$\bar{A}$	closure of A,
$\mathring{A}$	interior of A,
int A	interior of A,

$\cap$ intersection,

$\cup$ union,

$\setminus$ set - theoretic difference,

$\oplus$ direct sum,

$:=$ definition sign (left hand side is defined by right hand side),

o little o - symbol (Landau's symbol),

■ end of proof,

$\mathrm{dist}(x,M) = \mathrm{infimum}\{\|x-y\| : y \in M\}$ for a vector x and a set of vectors M,

$\mathrm{diag}(M_1,\ldots,M_k)$ for square matrices $M_1,\ldots,M_k$: the block diagonal matrix with the blocks $M_1,\ldots,M_k$ along the main diagonal,

$\mathrm{diag}(x)$ for a vector $x = (x_1,\ldots,x_n)$: the diagonal matrix with the elements $x_1,\ldots,x_n$ on the main diagonal.

A *principal fundamental matrix* $\Phi(t,s)$ of a linear system of differential equations or difference equations, respectively, is the fundamental matrix solution with $\Phi(s,s)$ = identity matrix.

A *trajectory, orbit, semitrajectory, semiorbit*, refers to a solution curve of an autonomous differential equation $\dot{x} = f(x)$ or difference equation $x(k+1) = f(x(k))$, respectively, in the x - space.

An *invariant manifold* is a manifold in x - space which is invariant with respect to the flow of an autonomous differential equation $\dot{x} = f(x)$ or difference equation $x(k+1) = f(x(k))$, respectively.

An *integral curve* is a curve $(t,x(t))$ where $x(t)$ is a solution of a nonautonomous differential equation $\dot{x} = f(t,x)$.

An *integral manifold* is a manifold in (t,x) - space which is invariant with respect to the integral curves of a nonautonomous differential equation $\dot{x} = f(t,x)$.

1. Introduction

The qualitative theory of differential equations or difference equations is mainly addressed to the various questions arising in the study of the long run behavior of solutions. The contents of this book is related to two of the major problems of the qualitative theory, namely stability and asymptotic behavior.

The stability problem is concerned with the behavior of solutions in a neighborhood of a given solution which may be taken to be an equilibrium of a nonautonomous equation. In this context most attention has been payed so far to the stability of isolated equilibria of differential equations and, much less, of difference equations. There are just a few results on how solutions behave near equilibrium points which are embedded in a continuum of equilibria. The reason for this is twofold. First, the appearance of a continuum of stationary solutions is exceptional when compared to isolated ones and, secondly, the case of non-isolated equilibria is much harder to deal with since the problems involved in this case are nonlocal. However, the study of continua of stationary solutions is well motivated, both from a theoretical and a practical point of view. On one hand, manifolds of stationary solutions represent simple examples of invariant manifolds or integral manifolds (see the notation section for the difference) and so results obtained for this case may give some insight into the general case. Moreover, some types of nonstationary manifolds can be reduced to stationary ones. On the other hand, several model equations from various disciplines such as physics, economy, biology and medicine possess manifolds of equilibria; one example from population genetics will be discussed in this book.

The asymptotic behavior of solutions can be determined by means of their ω - limit sets. Standard tools for locating those sets are Ljapunov functions. For instance, a first idea about the location of the ω - limit set of a particular solution can be obtained from LaSalle's Theorem. It is often necessary to analyze the properties of such an ω - limit set in more detail. To a large extent this book is devoted to the problem of finding sufficient conditions for an ω - limit set to be a singleton.

The general convergence problem which is treated in this text can be phrased as follows. What conditions have to be imposed on a solution $x(t)$ (of a differential equation or a difference equation) and on a continuum M of stationary solutions in order to guarantee the convergence of $x(t)$ to some point on M? In order to describe in some detail the contents of this book we now state kind of a prototype result (see Aulbach [5, Theorem 1]) which, later on, will be generalized, modified and applied to a biological problem.

<u>Proposition 1.1</u>: Let $x(t)$ be any solution of an autonomous differential equation

$$\dot{x} = f(x), \tag{1.1}$$

$f \in C^3(\mathbb{R}^n, \mathbb{R}^n)$, which has an m - dimensional differentiable manifold

M of stationary solutions. Suppose that

(i) the ω - limit set of $x(t)$ is not empty,

(ii) the distance between $x(t)$ and M decays to zero as $t \to \infty$,

(iii) for each $\bar{x} \in M$ $n-m$ eigenvalues of the Jacobian of f at $\bar{x}$ have real parts different from zero.

Then the solution $x(t)$ converges as $t \to \infty$ to some point on M.

Since we are interested in various modifications of this theorem the question arises if each of the three hypotheses (i), (ii), (iii) is needed for this convergence result. By means of examples we first show that this theorem becomes false if either one of the three hypotheses is dropped. In each example M will be an analytic manifold of dimension one.

Example 1: Consider the plane autonomous system

$$\dot{u} = u v^2$$
$$\dot{v} = -\frac{v}{u}$$

in the half-plane $u > 0$. The manifold M is the positive u-axis and each point $(\bar{u}, 0)$ on it has the eigenvalues 0 and $-1/\bar{u}$. As $x(t)$ we choose any solution whose trajectory $v = (2/u)^{1/2}$ solves the corresponding scalar equation $dv/du = -u^{-2} v^{-1}$. $x(t)$ approaches the manifold M as $t \to \infty$ without converging to a point on M. Hypotheses (ii) and (iii) are satisfied whereas (i) is not.

Example 2: The three - dimensional system (see Chow and Hale [10, p.360])

$$\dot{u} = 2v - \mu(u^3 - 6u^2 + v^2)(12u - 3u^2)$$
$$\dot{v} = 12u - 3u^2 + \mu(u^3 - 6u^2 + v^2)2v$$
$$\dot{w} = 0 ,$$

$\mu \in \mathbb{R}$, has the w-axis as manifold M of stationary solutions. For $\mu < 0$ the function $V(u,v) := u^3 - 6u^2 + v^2$ is a Ljapunov function which shows the following. In any hyperplane $w = c =$ constant there exists an homoclinic orbit contained in $\Gamma_c := \{(u,v,w): v^2 = 6u^2 - u^3, u \geq 0\}$ and any nonconstant solution $x_c(t)$ on $w = c$ inside this homoclinic orbit spirals outward onto Γ_c as $t \to \infty$. Thus, the ω-limit set Ω_c of $x_c(t)$ coincides with Γ_c. Assumptions (i) (with $\Omega_c \cap M = \{(0,0,c)\}$) and (iii) are satisfied and (ii) is not; $x_c(t)$ does not converge.

Example 3: In the plane polar coordinate system

$$\dot{r} = (1 - r)^3$$
$$\dot{\varphi} = (1 - r)^2$$

we take the unit circle $r = 1$ as manifold M of equilibria. The trajectories $1 + (r_o - 1)e^{-\varphi}$, $r_o > 0$, spiral onto M, thus satisfying (i) and (ii), without converging to any point on M. The eigenvalue condition (iii) is violated since every point on M has zero as double eigenvalue.

Having seen that the above result is false if either one of the three hypotheses is dropped one may ask whether there are less restrictive conditions which are sufficient for the convergence of $x(t)$. The results in this book will be derived under conditions which are weaker then those above in three respects. First, the global assumption (ii) will be replaced by a local one. Secondly, instead of requiring that the distance between $x(t)$ and M decays to zero we will only suppose that this distance remains sufficiently small. In the third modification we will replace the basic assumption that M carries a stationary flow by a certain stability assumption on the flow on M.

The organization of this book is as follows. In two parts we treat the continuous time case and the discrete time case, respectively, in a parallel manner. Part I starts with Section 2 where the basic result on the continuous time case is proved. The underlying equations are nonautonomous differential equations which need to be defined only for $t \geq 0$ and which have a stationary manifold M. In Section 3 the problem of the asymptotic phase is treated, i.e. the problem under what conditions a solution picks some point on M as its limit. Section 4 is devoted to the question whether it is possible for a solution to stay close to M without being on one of the stable manifolds which are associated with the family of equilibria on M. In Section 5 we discard the assumption of stationarity of the flow on M. Instead, we ask for a certain kind of stability of the flow on M. In Section 6 we apply the basic result of Section 2 to the classical selection model from population genetics.

Part II of this book is devoted to discrete time systems, to wit, to autonomous difference equations. The organization of this part is similar to part I, the techniques and the quoted results, however, differ from the continuous time case to a large extent. In any case additional hypotheses are necessary; first, to compensate for the fact that discrete time orbits are not connected whereas continuous time orbits are; secondly, to exclude the problems that could arise from a nonunique backward continuation of discrete time orbits. Section 7 contains the fundamental result of this work for difference equations where we assume that there is a stationary manifold M and a solution $x(t)$ whose distance from M decays to zero. In Section 8 we show that any solution $x(t)$ is asymptotic to some point on M under the weakened assumption that x(t) eventually stays sufficiently close to M. Section 9 demonstrates that a solution can remain near M for all future only if it decays on one member of the family of stable manifolds to the corresponding equilibrium on M. In Section 10 we weaken the assumption on the flow on M by just requiring that the solutions on M are stable in both time directions. Finally, Section 11 gives an application to the discrete time version of the basic selection model from population genetics.

The text is concluded by three appendices. In Appendix A we generalize a reducibility result due to Coppel for linear time varying differential systems. Appendix B provides some results on linear differential equations and difference equations which are crucial for the type of linearization analysis carried out in this book. Finally, in Appendix C we determine the solution set of a system of algebraic equations which gives all stationary solutions of the selection models treated in Sections 6 and 11.

In order to relate our work to the known literature we quote the few results on the asymptotic behavior of solutions of differential equations or difference equations near continua of stationary solutions. The first is due to Ljapunov [34, §31] (see also Malkin [35, §34]). Ljapunov showed that any solution of an autonomous differential equation which comes once sufficiently close to an asymptotically stable curve M of equilibria converges to some point on M. This corresponds to the case where in condition (iii) above m equals 1 and, for each $\bar{x} \in M$, $n-1$ eigenvalues of the Jacobian at $\bar{x}$ have negative real parts. For abstract evolution equations with an m-dimensional asymptotically stable manifold M this result has recently been proved by Henry [26, p.108]. From today's point of view the case of an asymptotically stable M is relatively easy because there exists a neighborhood of M which is invariantly fibred by stable manifolds. In the hyperbolic case the situation is much more complicated since in general there is no invariant fibration of any neighborhood of M (see Knobloch and Aulbach [31, p.181]). For this hyperbolic case Fenichel [18, Theorem 9.1(i)] gives an eight-line-sketch of a proof (see [18, p.97]) for a result which is in the spirit of our Section 3; Hale and Massatt [22, Theorem 2.1] state a result of the above type for a class of differential equations in a Banach space and indicate how it could be proved. For ordinary differential equations a complete proof of a convergence result like the proposition above in the general hyperbolic case can be found in Aulbach [5].

Also some results from invariant manifold theory are related to the subject of this book since, after all, a manifold of stationary solutions is an invariant manifold; in particular, under our assumptions on the eigenvalues it is a center manifold through each of its points. So, if a solution $x(t)$ eventually stays close enough to some point $\bar{x}$ on M it lies on a center - stable manifold through $\bar{x}$ and thus decays exponentially to some solution on M by the reduction principle (see e.g. Carr [8, Theorem 2], Aulbach [4, Theorem 4]). At first glance, this argument which is of a purely local nature seems to settle the above convergence question. However, our problem is a nonlocal one because the solution $x(t)$ is only supposed to stay near the manifold M; therefore it is allowed to leave small neighborhoods of any of the points on M again and again. The reduction of the global to the local situation is an idea of proof which is explicitly used in Sections 3 and 8. On the other hand, there is a basic global result on invariant manifolds due to Hirsch, Pugh and Shub [27, Theorem 4.1] which can be shown to imply results of the above type. This theorem applies both to diffeomorphisms and flows. However, results of this form (being derived by differential topological means) are not applicable in case the solutions of the underlying equation are not unique or even undefined in one time direction. Equations of this very type will be considered throughout this book, namely nonautonomous differential equations which are defined only for $t \geq 0$ (without obvious backward continuation due to the time variant linearization) and difference equations whose right hand sides are no diffeomorphisms. That this latter case is not only of theoretical interest will be demonstrated by the application to the selection model where just the non - diffeomorphism case occurs (see Section 11).

The feature of this book is that the equations we consider are of a form that there is no natural way of assigning to them one - parameter groups of transformations of the corresponding phase space. Thus, the general theory of dynamical systems and particularly the invariant manifold theory for diffeomorphisms and flows is not applicable. Under assumptions as weak as ours there are virtually no results in the literature on the general hyperbolic convergence problem, hence all of our theorems are innovative. Also the results we obtain for the basic selection models are novelties. In the continuous time case our results generalize those of Aulbach and Hadeler [7] because here the nonnegative viability parameters are not all supposed to be positive (as they are in [7]). For the discrete time model our result provides a solution of the convergence problem which has been open so far in arbitrary dimensions.

I. DIFFERENTIAL EQUATIONS

2. Basic Results

In this section we state and prove the basic result of this book for the continuous time case. The underlying equation is either autonomous or nonautonomous. We are interested in the asymptotic behavior of the solutions as the independent variable tends to infinity. So, in the general nonautonomous case, the equation must be defined on an interval which is unbounded to the right. In negative time direction our equation

$$\dot{x} = f(t,x) \tag{2.1}$$

need not even be defined. We suppose that the function f is continuous on $\mathbb{R}^+ \times \mathbb{R}^n$ and of class C^3 as function of x. In order to get the frame for our investigations we assume that

(I) equation (2.1) has a stationary solution x^* for all $t \geq 0$, i.e. $f(t,x^*) = 0$ for all $t \geq 0$,

(II) $\lim_{x \to x^*} f(t,x) = 0$ uniformly in t,

(III) there exists an m-dimensional C^1-manifold $M \subset \mathbb{R}^n$ of stationary solutions of (2.1) with $x^* \in M$, i.e. $f(t,x) = 0$ for all $(t,x) \in \mathbb{R}^+ \times M$.

At first glance it might look somewhat artificial to suppose that a nonautonomous differential equation has a manifold of stationary solutions. This situation, however, can be understood as a prototype for other situations which can be reduced to this one. Such a situation occurs e.g. if an autonomous differential equation has a family of periodic solutions (see Aulbach [1],[2]) as it frequently happens in Hamiltonian or reversible systems. Even certain types of nonperiodic invariant manifolds of autonomous equations can be reduced to stationary integral manifolds of nonautonomous equations (see Aulbach [3]).

Our first theorem is the core of the differential equation part of this book. The idea of its proof serves as a basis for the derivation of the results in Sections 3,4 and 5.

<u>Theorem 2.1</u>: Let $x(t)$ be any solution of (2.1) which exists for all $t \geq 0$. Denote its ω-limit set by Ω. Assume that

(i) $x^* \in \Omega$,

(ii) there exists an $\mathbb{R}^n$-neighborhood U of x^* such that $\Omega \cap U \subset M$,

(iii) the matrix $f_x(t,x^*)$ is bounded on $\mathbb{R}^+$ and admits an exponential dichotomy of the following type: there exist projections P^-, P^+ of $\mathbb{R}^n$ such that

$$\operatorname{rank} P^- + \operatorname{rank} P^+ = n - m$$

and positive constants α, β and γ with the following properties: a fundamental matrix $\Phi(t)$ of

$$\dot{y} = f_x(t,x^*)y$$

satisfies the estimates

$$\| \Phi(t)P^-\Phi^{-1}(s) \| \le \gamma e^{-\alpha(t-s)} \quad \text{for all } t \ge s \ge 0,$$

$$\| \Phi(t)P^+\Phi^{-1}(s) \| \le \gamma e^{-\beta(s-t)} \quad \text{for all } s \ge t \ge 0.$$

Then $\lim_{t \to \infty} x(t) = x^*$.

Proof: With the transformation $y = x - x^*$ we turn from equation (2.1) to the equation for the perturbed motion

$$\dot{y} = f_x(t,x^*)y + r(t,y) \tag{2.2}$$

where $r(t,y) = o(\| y \|)$ uniformly in t as $\| y \| \to 0$. Because of the dichotomy assumption the linearization of (2.2) satisfies the hypotheses of Theorem A.1, Appendix A. The matrix $T(t)$ of this theorem is used to transform system (2.2) via

$$y = T(t)\begin{pmatrix} u \\ v \\ w \end{pmatrix}, \quad u \in \mathbb{R}^{n_-}, \quad v \in \mathbb{R}^{n_+}, \quad w \in \mathbb{R}^m \tag{2.3}$$

where $n_- := \operatorname{rank} P^-$, $n_+ := \operatorname{rank} P^+$, into a system of the form

$$\begin{aligned} \dot{u} &= A^-(t)u + r_1(t,u,v,w) \\ \dot{v} &= A^+(t)v + r_2(t,u,v,w) \\ \dot{w} &= A^o(t)w + r_3(t,u,v,w) \end{aligned} \tag{2.4}$$

where the principal fundamental matrices $\Phi^-(t,s)$, $\Phi^+(t,s)$ of $\dot{u} = A^-(t)u$, $\dot{v} = A^+(t)v$, respectively, satisfy the estimates

$$\begin{aligned} \| \Phi^-(t,s) \| &\le \tilde{\gamma} e^{-\alpha(t-s)} \quad \text{for all } t \ge s \ge 0, \\ \| \Phi^+(t,s) \| &\le \tilde{\gamma} e^{-\beta(s-t)} \quad \text{for all } s \ge t \ge 0, \end{aligned} \tag{2.5}$$

$\tilde{\gamma}$ being a positive constant. Moreover, the nonlinearities r_1, r_2, r_3 are continuous functions which are of class C^2 as functions of u, v, w and of order $o(\|(u,v,w)\|)$ uniformly in t as $\|(u,v,w)\| \to 0$. Here we have used (and later on we shall frequently use without further notice) that the matrix T(t) together with its inverse is continuously differentiable and bounded on $\mathbb{R}^+$. When we denote by $X: H \to \mathbb{R}^n$, H a neighborhood of $0 \in \mathbb{R}^m$, a local C^1-coordinate system of the manifold M of stationary solutions of (2.1) near x^*, $X(0) = x^*$, then

$$(U(t,\eta), V(t,\eta), W(t,\eta)) := T^{-1}(t)\,[X(\eta) - x^*] \tag{2.6}$$

represents an m-parameter family of solutions of (2.4) with

$$(U(t,0), V(t,0), W(t,0)) \equiv (0,0,0) \text{ on } \mathbb{R}^+.$$

Furthermore, the m functions

$$\begin{aligned} &(U_{\eta^i}(t,0), V_{\eta^i}(t,0), W_{\eta^i}(t,0)) = \\ &T^{-1}(t)\, X_{\eta^i}(0), \quad i = 1,\dots,m, \end{aligned} \tag{2.7}$$

are linearly independent solutions of the linearization

$$\begin{aligned} \dot{u} &= A^-(t)u \\ \dot{v} &= A^+(t)v \\ \dot{w} &= A^o(t)w \end{aligned} \tag{2.8}$$

of (2.4). This means, in particular, that the relation

$$\begin{aligned} u &= U(t,\bar{w}) + \bar{u} \\ v &= V(t,\bar{w}) + \bar{v} \\ w &= W(t,\bar{w}) \end{aligned} \tag{2.9}$$

represents a local transformation in a tube around the t-axis in the (t,u,v,w)-space. Introducing the abbreviations

$$\begin{aligned} r_i^*(t,\bar{u},\bar{v},\bar{w}) &:= r_i(t,\bar{u}+U(t,\bar{w}),\bar{v}+V(t,\bar{w}),W(t,\bar{w})) - \\ & r_i(t,U(t,\bar{w}),V(t,\bar{w}),W(t,\bar{w})), \quad i=1,2,3, \end{aligned} \tag{2.10}$$

we get the $(\bar{u},\bar{v},\bar{w})$-system in the form

$$\begin{aligned} \dot{\bar{u}} &= A^-(t)\bar{u} + r_1^*(t,\bar{u},\bar{v},\bar{w}) - U_\eta(t,\bar{w})\, W_w^{-1}(t,W(t,\bar{w}))\, r_3^*(t,\bar{u},\bar{v},\bar{w}) \\ \dot{\bar{v}} &= A^+(t)\bar{v} + r_2^*(t,\bar{u},\bar{v},\bar{w}) - V_\eta(t,\bar{w})\, W_w^{-1}(t,W(t,\bar{w}))\, r_3^*(t,\bar{u},\bar{v},\bar{w}) \\ \dot{\bar{w}} &= W_w^{-1}(t,W(t,\bar{w}))\, r_3^*(t,\bar{u},\bar{v},\bar{w}) \end{aligned}$$

where $W_w^{-1}(t,\cdot)$ denotes the Jacobian of the inverse function $W^{-1}(t,\cdot)$ of $W(t,\cdot)$. Obviously, this transformed system may be written as

$$\begin{aligned} \dot{\bar{u}} &= A^-(t)\bar{u} + \bar{r}_1(t,\bar{u},\bar{v},\bar{w}) \\ \dot{\bar{v}} &= A^+(t)\bar{v} + \bar{r}_2(t,\bar{u},\bar{v},\bar{w}) \\ \dot{\bar{w}} &= \bar{r}_3(t,\bar{u},\bar{v},\bar{w}) \end{aligned} \tag{2.11}$$

where the nonlinearities are continuous, of class C^1 as functions of $\bar{u}$, $\bar{v}$, $\bar{w}$ and of order $o(\|(\bar{u},\bar{v},\bar{w})\|)$ uniformly in t as $\|(\bar{u},\bar{v},\bar{w})\| \to 0$. Moreover, they satisfy the identities $\bar{r}_1(t,0,0,\bar{w}) = 0$, $\bar{r}_2(t,0,0,\bar{w}) = 0$, $\bar{r}_3(t,0,0,\bar{w}) = 0$ for all $t \in \mathbb{R}^+$ and $\|\bar{w}\|$ suitably small. This just means that (2.11) admits the local family $(0,0,\bar{w})$ of stationary solutions.

In order to obtain for system (2.11) a certain kind of decoupling (see system (2.22) below) we have to derive now the existence of a certain integral manifold for system (2.11). The main tool we are using in the construction of this manifold is a theorem due to Knobloch which we state for completeness.

Theorem 2.2 (Knobloch and Kappel [30, V.Satz 7.1]): Consider the differential system

$$\dot{z} = p(t,z,v)$$
$$\dot{v} = A^{+}(t)v + q(t,z,v) \tag{2.12}$$

where $t \in \mathbb{R}^{+}$, $z \in \mathbb{R}^{m+n_-}$, $v \in \mathbb{R}^{n_+}$, under the following hypotheses:

(1) Smoothness: $(p,q) \in C(\mathbb{R}^{+} \times \mathbb{R}^{n}, \mathbb{R}^{n})$; $(p(t,\cdot,\cdot), q(t,\cdot,\cdot)) \in C^{2}(\mathbb{R}^{n}, \mathbb{R}^{n})$,

(2) Global boundedness: p, q and their derivatives are bounded throughout $\mathbb{R}^{+} \times \mathbb{R}^{n}$,

(3) Exponential dichotomy: There exist positive constants γ_1, γ_2, ζ and $\delta \in \mathbb{R}$ with $\delta < \zeta$ having the following properties: if $(z(t), v(t))$ is any solution of (2.12) then the principal fundamental matrices $\Psi_1(t,s)$, $\Psi_2(t,s)$ of $\dot{z} = p_z(t,z(t),v(t))z$, $\dot{v} = [A^{+}(t) + q_v(t,z(t), v(t))]v$, respectively, obey the estimates

$$\| \Psi_1(t,s) \| \leq \gamma_1 e^{\delta(t-s)} \quad \text{for all } t \geq s \geq 0,$$
$$\| \Psi_2(t,s) \| \leq \gamma_2 e^{\zeta(t-s)} \quad \text{for all } s \geq t \geq 0.$$

(4) Weak coupling: There exists a $\rho \geq 0$ such that for $\nu = 1,2$

$$\delta < \nu\rho < \zeta$$

and

$$\kappa_{\rho,\nu} := \gamma_1\gamma_2 \sup \| p_v(t,z,v) \| \sup \| q_z(t,z,v) \| (\nu\rho-\delta)^{-1} (\zeta-\nu\rho)^{-1} < 1,$$

where in either case the supremum is taken over all $(t,z,v) \in \mathbb{R}^{+} \times \mathbb{R}^{n}$.

Under these hypotheses there exists a C^1 - function $v = s(t,z)$ with a bounded z - derivative having the following properties:

(i) The set $S := \{(t,z,v): v = s(t,z),\ (t,z) \in \mathbb{R}^+ \times \mathbb{R}^{m+n_-}\}$ is the maximal integral manifold of solutions of (2.12) whose v-components are bounded as $t \to \infty$.

(ii) The identity

$$s_t(t,z) + s_z(t,z)\, p(t,z,s(t,z)) = A^+(t)s(t,z) + q(t,z,s(t,z)) \qquad (2.13)$$

holds for all $(t,z) \in \mathbb{R}^+ \times \mathbb{R}^{m+n_-}$.

(iii) For all $(t,z) \in \mathbb{R}^+ \times \mathbb{R}^{m+n_-}$ the following estimate holds true where each supremum is taken over all $(t,z,v) \in \mathbb{R}^+ \times \mathbb{R}^n$:

$$\| s_z(t,z) \| \leq (1-\kappa_{\rho,1})^{-1} \gamma_2 (\zeta - \rho)^{-1} \sup \| q_z(t,z,v) \| \times$$
$$[1 + \max\{\sup \| p_z(t,z,v) \|\ ,\ \sup \| A^+(t) + q_v(t,z,v) \| \} \gamma_1 (\rho - \delta)^{-1}].$$

This theorem differs slightly from Knobloch's original one in two respects. First, our conditions on system (2.12) as well as the conclusions are supposed to hold only for $t \geq 0$ rather than for all $t \in \mathbb{R}$. The reduction of our situation to the original one can easily be accomplished by setting $p(t,z,v) := p(0,z,v)$, $q(t,z,v) := q(0,z,v)$, $A^+(t) := A^+(0)$ for all $t \in (-\infty,0]$. Secondly, in the original theorem in [30] A^+ is a constant matrix whose eigenvalues have positive real parts. The proof in [30], however, does not use this fact explicitly. Since only estimates for the corresponding principal fundamental matrix appear our assumptions on the matrix $A^+(t)$ are sufficient for the original proof to carry over to our situation.

Let us return to system (2.11). The existence of a time variant "center - stable manifold" $\bar{v} = s(t,\bar{u},\bar{w})$ for this system now follows quite easily from Theorem 2.2.

Lemma 2.1: There exists a positive number r_o such that system (2.11) has a local integral manifold

$$\{(t,\bar{u},\bar{v},\bar{w}):\ \bar{v} = s(t,\bar{u},\bar{w}),\ t \in \mathbb{R}^+,\ \|(\bar{u},\bar{v})\| < r_o\}$$

where s is of class C^1 and has the following properties:

(i) $s(t,\bar{u},\bar{w})$ satisfies for all $t \geq 0$ and $\|(\bar{u},\bar{v})\| < r_o$ the partial differential equation

$$\begin{aligned} &s_t(t,\bar{u},\bar{w}) + s_{\bar{u}}(t,\bar{u},\bar{w})[A^-(t)\bar{u} + \bar{r}_1(t,\bar{u},s(t,\bar{u},\bar{w}),\bar{w})] + \\ &s_{\bar{w}}(t,\bar{u},\bar{w})[\bar{r}_3(t,\bar{u},s(t,\bar{u},\bar{w}),\bar{w})] = \\ &A^+(t)s(t,\bar{u},\bar{w}) + \bar{r}_2(t,\bar{u},s(t,\bar{u},\bar{w}),\bar{w}). \end{aligned} \tag{2.14}$$

(ii) $s(t,0,\bar{w}) = 0$, $s_{(\bar{u},\bar{w})}(t,0,0) = 0$ for all $t \in \mathbb{R}^+$ and $\|\bar{w}\|$ suitably small.

Proof: In the well known way (see e.g. Knobloch and Kappel [30]) we replace system (2.11) by a system of the form

$$\begin{aligned} \dot{\bar{z}} &= A(t)\bar{z} + \tilde{r}(t,\bar{z},\bar{v}) \\ \dot{\bar{v}} &= A^+(t)\bar{v} + \tilde{r}_3(t,\bar{z},\bar{v}) \end{aligned}$$

($\bar{z} = (\bar{u},\bar{v})$, $A(t) := \mathrm{diag}(A^-(t),O_{m\times m})$) which coincides with (2.11) on a set of the form $\{(t,\bar{z},\bar{v}):\ t \geq 0,\ \|(\bar{z},\bar{v})\| < r\}$ and moreover satisfies the global boundedness requirements of Theorem 2.2. By an appropriate choice of r we can make these global bounds as small as desired and we thus can verify (see Lemma B.1, Appendix B) the dichotomy condition (3) of Theorem 2.2 with $\zeta = \alpha - \varepsilon$, $\delta = \varepsilon$, where the positive number ε can be made as small as desired by a suitable choice of r. Note here that because of the particular form of $A(t)$ the principal fundamental matrix $\bar{\Phi}(t,s)$ of $\dot{\bar{z}} = A(t)\bar{z}$ satisfies, for some positive $\bar{\gamma}$, the estimate

$$\| \bar{\Phi}(t,s) \| \leqq \bar{\gamma} \quad \text{for all } t \geqq s \geqq 0. \tag{2.15}$$

Hence, we may take as ρ in the coupling condition (4) any number in the open interval $(0,\alpha)$. Again, by choosing r small we verify condition (4) of Theorem 2.2 and thus its full set of hypotheses. This theorem provides then a local integral manifold $\bar{v} = s(t,\bar{u},\bar{w})$ for system (2.11) where s has the desired smoothness properties and obviously satisfies the partial differential equation (2.14).

Next we prove that the function $S(t) := s_{\bar{z}}(t,0,0)$ vanishes identically on $\mathbb{R}^+$. First of all, note that the elements of the matrix $S(t)$ are differentiable on $\mathbb{R}^+$ and, because of (iii) of Theorem 2.2, that

$$S(t) \text{ is bounded on } \mathbb{R}^+. \tag{2.16}$$

Inserting the expression $s(t,\bar{z}) = S(t)\bar{z} + o(\|\bar{z}\|)$ into equation (2.14) and equating coefficients of like powers of $\bar{z}$ we get

$$\dot{S}(t)\bar{z} + S(t)A(t)\bar{z} = A^+(t)S(t)\bar{z}$$

for all $t \geqq 0$ and $\|\bar{z}\| < r_o$. This means that $\bar{v} = S(t)\bar{z}$ describes a global linear integral manifold of the decoupled linear system

$$\begin{aligned} \dot{\bar{z}} &= A(t)\bar{z} \\ \dot{\bar{v}} &= A^+(t)\bar{v}. \end{aligned} \tag{2.17}$$

Now, in order to prove the identity $S(t) \equiv 0$ on $\mathbb{R}^+$ we suppose to the contrary that there exists a $t_o \geq 0$ and a $\bar{z}_o \in \mathbb{R}^{m+n_-}$ such that

$$S(t_o)\bar{z}_o \neq 0. \tag{2.18}$$

Let $\bar{z}(t,t_o,\bar{z}_o)$ denote the solution of $\dot{\bar{z}} = A(t)\bar{z}$ with initial value $\bar{z}_o$ for $t = t_o$. From (2.15) it is apparent that

$$\| \bar{z}(t,t_o,\bar{z}_o) \| \leq \bar{\gamma} \, \| \bar{z}_o \| \quad \text{for all } t \geq t_o. \tag{2.19}$$

Since $\bar{v} = S(t)\bar{z}$ is an integral manifold for (2.17) the function $S(t)\bar{z}(t,t_o,\bar{z}_o)$ is a solution of $\dot{\bar{v}} = A^+(t)\bar{v}$, i.e.

$$S(t)\bar{z}(t,t_o,\bar{z}_o) = \Phi^+(t,t_o)S(t_o)\bar{z}_o,$$

or equivalently,

$$\Phi^+(t_o,t)S(t)\bar{z}(t,t_o,\bar{z}_o) = S(t_o)\bar{z}_o \quad \text{for all } t \geq t_o.$$

With the corresponding estimate in (2.5) this leads to the inequality

$$\| S(t_o)\bar{z}_o \| \leq \| S(t) \| \, \| \bar{z}(t,t_o,\bar{z}_o) \| \, \tilde{\gamma} \, e^{-\beta(t-t_o)} \quad \text{for all } t \geq t_o$$

which in turn yields with (2.19)

$$\| S(t_o)\bar{z}_o \| \leq \| S(t) \| \, \bar{\gamma} \, \|\bar{z}_o\| \, \tilde{\gamma} \, e^{-\beta(t-t_o)} \quad \text{for all } t \geq t_o. \tag{2.20}$$

Since $\| S(t) \|$ is bounded on $\mathbb{R}^+$ (see (2.16)) the right hand side of (2.20) tends to zero as $t \to \infty$. Thus, we get the relation $\| S(t_o)\bar{z}_o \| = 0$ and herewith a contradiction to (2.18).

After having proved that $s_{\bar{z}}(t,0,0)$ is identically zero it remains to be shown that $s(t,0,\bar{w})$ vanishes for all $t \in \mathbb{R}^+$ and $\|\bar{w}\|$ small. This, however, follows immediately from the maximality of s (see (i) in Theorem 2.2) and the fact that equation (2.11) admits the family $(0,0,\bar{w})$, $\|\bar{w}\|$ small, of stationary solutions. This completes the proof of Lemma 2.1. ∎

Having the function $s(t,\bar{u},\bar{w})$ at hand we continue the proof of Theorem 2.1. We apply to system (2.11) the transformation

$$\hat{u} = \bar{u}$$
$$\hat{v} = \bar{v} - s(t,\bar{u},\bar{w})$$
$$\hat{w} = \bar{w}.$$

The crucial $\hat{v}$-equation turns out to be

$$\begin{aligned}\dot{\hat{v}} = A^+(t)\bar{v} + \bar{r}_2(t,\bar{u},\bar{v},\bar{w}) - s_t(t,\hat{u},\hat{w}) - \\ s_{\bar{u}}(t,\hat{u},\hat{w})[A^-(t)\hat{u} + \bar{r}_1(t,\hat{u},\hat{v}+s(t,\hat{u},\hat{w}),\hat{w})] - \\ s_{\bar{w}}(t,\hat{u},\hat{w})[\bar{r}_3(t,\hat{u},\hat{v}+s(t,\hat{u},\hat{w}),\hat{w})],\end{aligned}$$

which after introducing the abbreviations

$$R_i(t,\hat{u},\hat{v},\hat{w}) := \bar{r}_i(t,\hat{u},\hat{v}+s(t,\hat{u},\hat{w}),\hat{w}) - \bar{r}_i(t,\hat{u},s(t,\hat{u},\hat{w}),\hat{w}),\ i=1,3,$$

and using (2.14) may be written as

$$\begin{aligned}\dot{\hat{v}} = A^+(t)\hat{v} + \bar{r}_2(t,\hat{u},\hat{v}+s(t,\hat{u},\hat{w}),\hat{w}) - \bar{r}_2(t,\hat{u},s(t,\hat{u},\hat{w}),\hat{w}) - \\ s_{\bar{u}}(t,\hat{u},\hat{w})\,R_1(t,\hat{u},\hat{v},\hat{w}) - s_{\bar{w}}(t,\hat{u},\hat{w})\,R_3(t,\hat{u},\hat{v},\hat{w}).\end{aligned}$$

Hence, we get the $(\hat{u},\hat{v},\hat{w})$-system in the form

$$\begin{aligned}\dot{\hat{u}} &= A^-(t)\hat{u} + \hat{r}_1(t,\hat{u},\hat{v},\hat{w}) \\ \dot{\hat{v}} &= A^+(t)\hat{v} + \hat{r}_2(t,\hat{u},\hat{v},\hat{w}) \\ \dot{\hat{w}} &= \hat{r}_3(t,\hat{u},\hat{v},\hat{w})\end{aligned} \tag{2.21}$$

possessing all properties of (2.11), in particular $\hat{r}_1(t,0,0,\hat{w}) = 0$, $\hat{r}_3(t,0,0,\hat{w}) = 0$ for all $t \geq 0$ and $\|\hat{w}\|$ small, and moreover $\hat{r}_2(t,\hat{u},0,\hat{w}) = 0$ for all $t \geq 0$, $\|\hat{u}\|$ and $\|\hat{w}\|$ suitably small. Thus, setting

$$A_i(t,\hat{u},\hat{v},\hat{w}) := \int_0^1 (\hat{r}_i)_{\hat{u}}(t,s\hat{u},s\hat{v},\hat{w})ds, \quad i = 1,3,$$

$$B_j(t,\hat{u},\hat{v},\hat{w}) := \int_0^1 (\hat{r}_j)_{\hat{v}}(t,s\hat{u},s\hat{v},\hat{w})ds, \quad j = 1,2,3,$$

we may write system (2.21) in the particular form

$$\begin{aligned}\dot{\hat{u}} &= A^-(t)\hat{u} + A_1(t,\hat{u},\hat{v},\hat{w})\hat{u} + B_1(t,\hat{u},\hat{v},\hat{w})\hat{v} \\ \dot{\hat{v}} &= A^+(t)\hat{v} + B_2(t,\hat{u},\hat{v},\hat{w})\hat{v} \\ \dot{\hat{w}} &= A_3(t,\hat{u},\hat{v},\hat{w})\hat{u} + B_3(t,\hat{u},\hat{v},\hat{w})\hat{v}\end{aligned} \tag{2.22}$$

where the matrices A_i, B_j are continuous functions of all their variables and tend to zero uniformly in $t \in \mathbb{R}^+$ as $\|(\hat{u},\hat{v},\hat{w})\| \to 0$. System (2.22) exhibits a certain kind of linearity and decoupling. Just those properties will now be used to complete the proof of Theorem 2.1.

Let us recall the situation of this theorem, but now in $(\hat{u},\hat{v},\hat{w})$-coordinates. There exists a solution $(\hat{u}(t),\hat{v}(t),\hat{w}(t))$ of (2.22) and a sequence $t_\nu \to \infty$ as $\nu \to \infty$ such that

$$\lim_{\nu\to\infty} (\hat{u}(t_\nu),\hat{v}(t_\nu),\hat{w}(t_\nu)) = (0,0,0). \tag{2.23}$$

Furthermore, there exists a σ-ball B_σ around $(0,0,0)$ in $\mathbb{R}^n$ with the property that for any ω - limit point $(\hat{u}_\infty,\hat{v}_\infty,\hat{w}_\infty)$ of $(\hat{u}(t),\hat{v}(t),\hat{w}(t))$ in B_σ we get

$$\hat{u}_\infty = 0, \ \hat{v}_\infty = 0. \tag{2.24}$$

It is the final goal of the proof to show that the ω - limit point (0,0,0) of the solution $(\hat{u}(t),\hat{v}(t),\hat{w}(t))$ of (2.22) is the limit of this solution as $t\to\infty$. In order to prove this we suppose the contrary. This means that there exists a small ball $B_{\hat{\sigma}}$, say, around (0,0,0) such that the solution $(\hat{u}(t),\hat{v}(t),\hat{w}(t))$ cannot remain in $B_{\hat{\sigma}}$ for all t in an interval of the form $[T,\infty)$. Since at each instance t_ν, $\nu \in \mathbb{N}_0$ (w.l.o.g.), the solution $(\hat{u}(t),\hat{v}(t),\hat{w}(t))$ is inside $B_{\hat{\sigma}}$, it must leave and enter $B_{\hat{\sigma}}$ again and again. We thus may assume that there exists a sequence of compact intervals $[t_\nu,T_\nu]$, $t_\nu < T_\nu$, such that

$$\| (\hat{u}(t),\hat{v}(t),\hat{w}(t)) \| < \hat{\sigma} \text{ for all } t \in [t_\nu,T_\nu), \tag{2.25}$$

$$\| (\hat{u}(T_\nu),\hat{v}(T_\nu),\hat{w}(T_\nu)) \| = \hat{\sigma} \text{ for all } \nu \in \mathbb{N}_0. \tag{2.26}$$

Without loss of generality we take $\hat{\sigma}$ so small that $\hat{\sigma} < \sigma$ and

$$\begin{aligned} &\| A_i(t,\hat{u}(t),\hat{v}(t),\hat{w}(t)) \| \le \frac{\min\{\alpha,\beta\}}{2\gamma}, \quad i = 1,3, \\ &\| B_j(t,\hat{u}(t),\hat{v}(t),\hat{w}(t)) \| \le \frac{\min\{\alpha,\beta\}}{2\gamma}, \quad j = 1,2,3, \end{aligned} \tag{2.27}$$

on each interval $[t_\nu,T_\nu]$, $\nu \in \mathbb{N}_0$. α,β and γ are the constants given in Theorem 2.1. Thus, for each $\nu \in \mathbb{N}_0$, $(\hat{u}(t),\hat{v}(t),\hat{w}(t))$ can be considered as a solution of the linear system

$$\begin{aligned} \dot{\hat{u}} &= A^-(t)\hat{u} + C_1(t)\hat{u} + D_1(t)\hat{v} \\ \dot{\hat{v}} &= A^+(t)\hat{v} \qquad\qquad\;\; + D_2(t)\hat{v} \\ \dot{\hat{w}} &= \qquad\qquad\;\; C_3(t)\hat{u} + D_3(t)\hat{v} \end{aligned} \tag{2.28}$$

on $[t_\nu, T_\nu]$ where

$$C_i(t) := A_i(t,\hat{u}(t),\hat{v}(t),\hat{w}(t)), \quad i = 1,3,$$
$$D_j(t) := B_j(t,\hat{u}(t),\hat{v}(t),\hat{w}(t)), \quad j = 1,2,3.$$

The linear system (2.28) satisfies the assumptions of Lemma B.2, Appendix B, and this provides the estimates

$$\|\hat{w}(T_\nu)\| \leq \|\hat{w}(t_\nu)\| + \|\hat{u}(t_\nu)\| + \tfrac{3}{2}\|\hat{v}(T_\nu)\| \quad \text{for all } \nu \in \mathbb{N}_0. \tag{2.29}$$

Since $(\hat{u}(T_\nu),\hat{v}(T_\nu),\hat{w}(T_\nu))$, $\nu \in \mathbb{N}_0$, is a bounded sequence there exists a convergent subsequence $(\hat{u}(T_{\nu_\mu}),\hat{v}(T_{\nu_\mu}),\hat{w}(T_{\nu_\mu}))$ with limit $(\hat{u}_\infty,\hat{v}_\infty,\hat{w}_\infty)$, say, in B_σ. With (2.24) and (2.26) we thus get

$$\lim_{\mu\to\infty} \hat{u}(T_{\nu_\mu}) = 0, \tag{2.30}$$

$$\lim_{\mu\to\infty} \hat{v}(T_{\nu_\mu}) = 0, \tag{2.31}$$

$$\lim_{\mu\to\infty} \|\hat{w}(T_{\nu_\mu})\| = \hat{\sigma} > 0. \tag{2.32}$$

The relation (2.31) together with (2.23) implies that the right hand side of (2.29), for $\nu = \nu_\mu \in \mathbb{N}_0$, tends to zero as $\mu \to \infty$. Then also the left hand side tends to zero, i.e.

$$\lim_{\mu\to\infty} \hat{w}(T_{\nu_\mu}) = 0.$$

This, however, is a contradiction to (2.32) and the proof of Theorem 2.1 is complete. ∎

As a corollary we state Theorem 2.1 as it appears for autonomous equations. This will be applied in Section 6 below to a model equation arising in population genetics.

<u>Theorem 2.3</u>: Consider an autonomous system

$$\dot{x} = f(x) \tag{2.33}$$

where f belongs to the class $C^3(\mathbb{R}^n, \mathbb{R}^n)$. Suppose that (2.33) has a differentiable manifold M of stationary solutions. Let x(t) be any solution of (2.33) with ω - limit set Ω. Suppose that

(i) there exists a point $x^* \in \Omega$, i.e. $\Omega \neq \emptyset$,

(ii) there exists a neighborhood U of x^* such that $\Omega \cap U \subset M$,

(iii) $n - \dim M$ eigenvalues of $f_x(x^*)$ have real parts $\neq 0$.

Then $\lim_{t \to \infty} x(t) = x^*$.

We note in passing that the above proof simplifies considerably in the situation of Theorem 2.3 because equation (2.33) defines a differentiable dynamical system and so the complete machinery of invariant manifold theory is available. In fact, instead of deriving the local normal form (2.22) near x^* one can use the topological equivalence of system (2.33) near x^* to its linearization around x^* (see Palmer [36],[37]). This, however, does not completely settle the problem since the homeomorphic image $H(x(t))$ (H is the homeomorphism establishing the topological equivalence) of the solution x(t) of Theorem 2.3 is by assumption not forbidden to leave again and again the small region where the topological equivalence is valid. So a priori one does not know that $H(x(t))$ is a solution of a linear system on an interval of the form $[t_0, \infty)$. That this is indeed the case can be shown by means of arguments resembling those of the previous proof.

Finally it should be mentioned that the situation of Theorem 2.3 (not that of Theorem 2.1!) can be put into the framework of global analysis. Roughly speaking M (if supposed to be compact) is normally hyperbolic and so it follows from Hirsch, Pugh and Shub [27, Theorem 4.1] that any solution of (2.33) which eventually stays close enough to M lies on the stable manifold associated with some point on M. This obviously implies the result of Theorem 2.3. Furthermore, it indicates that the assumption of x(t) having its ω - limit points on M may be weakened to the assumption that the ω - limit set of x(t) lies in a sufficiently small neighborhood of M. Such a result is also true in the general setting of our work which is not amenable to global analysis results. This is the content of the next two sections.

3. Asymptotic Phase

In the previous section we assumed that a solution x(t) has an ω - limit point on the manifold M and we concluded that, under certain conditions, this ω - limit point actually is the limit of x(t) as $t \to \infty$. In this section we show under what conditions a solution is able to pick a point on M as its limit. This is the problem of the existence of an asymptotic phase.

The notion *asymptotic phase* first appeared in connection with orbitally asymptotically stable periodic solutions of autonomous differential equations (see Coddington and Levinson [11, Chapter 13, Theorem 2.2], Cesari [9, p.99]). Later it has been extended to more general invariant sets than closed orbits; see Fenichel [16, I.E.], Henry [26, Corollary 6.1.5] and Aulbach [3], the latter also for more references. Roughly speaking, an *asymptotic phase* associated with some invariant set Q is a map φ from some set S into Q with the property that for any $x \in S$ the difference $X(t,x) - X(t,\varphi(x))$ tends to zero as $t \to \infty$, where X represents the underlying dynamical system. Local center manifolds around a stable equilibrium are rather general types of invariant manifolds with asymptotic phase. For autonomous equations see Carr [8, Theorem 2], for nonautonomous equations and weakened stability assumptions see Aulbach [4, Theorem 4]. If the invariant set Q is asymptotically stable and has enough regularity, then an asymptotic phase exists and is a smooth map on S (see Fenichel [17, Theorem 5] and Aulbach [4, Theorem 4]). Until only recently the occurrence of an asymptotic phase was always connected with asymptotic stability of the corresponding invariant set. That an asymptotic phase can exist also without the assumption of asymptotic stability has been shown in Aulbach [1] and [2, Theorem 4] for hyperbolic manifolds of periodic solutions.

This book is devoted to manifolds of stationary solutions and so we further restrict attention to this case. Approaching a set Q of stationary solutions with asymptotic phase then simply means converging to some point on Q. Asymptotically stable manifolds have been treated by Ljapunov [34, §31] for autonomous ordinary differential equations and by Henry [26, p.108] for abstract evolution equations. The hyperbolic case has been considered by Fenichel [18, Theorem 9.1] and Aulbach [5] for autonomous ordinary differential equations and by Hale and Massatt [22, Theorem 2.1] for a class of partial differential equations.

In this section we deal with nonautonomous ordinary differential equations which, as in the previous section, are not necessarily defined on the entire real line. We consider equations of the form

$$\dot{x} = f(t,x) \tag{3.1}$$

where f is supposed to be continuous on $\mathbb{R}^+ \times \mathbb{R}^n$ and of class C^3 as a function of x. Furthermore we suppose that

(i) system (3.1) has a compact m-dimensional C^1-manifold $M \subset \mathbb{R}^n$ of stationary solutions,

(ii) for each $\bar{x} \in M$

$$\lim_{x \to \bar{x}} f(t,x) = 0 \quad \text{uniformly in } t,$$

(iii) for each $\bar{x} \in M$ there exist projections P^-, P^+ of $\mathbb{R}^n$ such that

$$\operatorname{rank} P^- + \operatorname{rank} P^+ = n - m,$$

and positive constants α, β and γ with the following properties: for a fundamental matrix $\Phi(t)$ of

$$\dot{y} = f_x(t,\bar{x})y \tag{3.2}$$

the following estimates hold true:

$$\| \Phi(t)P^-\Phi^{-1}(s) \| \leq \gamma e^{-\alpha(t-s)} \quad \text{for all } t \geq s \geq 0,$$

$$\| \Phi(t)P^+\Phi^{-1}(s) \| \leq \gamma e^{-\beta(s-t)} \quad \text{for all } s \geq t \geq 0.$$

Remarks: 1. Note that M is a manifold in the phase space $\mathbb{R}^n$ whereas $\mathbb{R}^+ \times M$ is the corresponding integral manifold in the extended phase space $\mathbb{R}^+ \times \mathbb{R}^n$.

2. In assumption (iii) the projections P^-, P^+ as well as the constants α, β and γ may depend upon $\bar{x}$.

If a solution $x(t)$ of (3.1) tends to M in the sense that

$$\lim_{t \to \infty} \operatorname{dist}(x(t), M) = 0,$$

then the compactness of M guarantees that there exists an ω - limit point x^* of $x(t)$ on M and by means of Theorem 2.1 we get

$$\lim_{t \to \infty} x(t) = x^*.$$

It is the aim of this section to prove a generalization of this result which can be useful from a perturbation point of view. The assumption that a given solution tends to the manifold M as $t \to \infty$ is replaced by the assumption that a solution remains sufficiently close to M as $t \to \infty$.

<u>Theorem 3.1</u>: Under the above hypotheses (i), (ii), (iii) there exists an $\mathbb{R}^n$- neighborhood N of M with the following property: if an integral curve $(t, x(t))$ of (3.1) eventually remains in $\mathbb{R}^+ \times N$, then there exists a point $x^* \in M$ such that

$$\lim_{t \to \infty} x(t) = x^*.$$

<u>Proof</u>: We suppose to the contrary that for any neighborhood $\mathbb{R}^+ \times \tilde{N}$ of $\mathbb{R}^+ \times M$ there exists a t_o and a solution $x_o(t)$ of (3.1) such that $(t, x_o(t))$ belongs to $\mathbb{R}^+ \times \tilde{N}$ for all $t \geq t_o$ and $x_o(t)$ does not converge to any point on M as $t \to \infty$. This particularly means that there exists a sequence of solutions $x_\nu(t)$ of (3.1) and a sequence of instants t_ν

such that

$$x_\nu(t) \in N_\nu \quad \text{for all } t \geq t_\nu \text{ where}$$
$$N_\nu := \{x \in \mathbb{R}^n : \operatorname{dist}(x,M) \leq \frac{1}{\nu}\}, \quad \nu \in \mathbb{N}_0, \tag{3.3}$$

and

$$x_\nu(t) \text{ does not converge to any point on } M \text{ as } t \to \infty. \tag{3.4}$$

The compactness of each N_ν guarantees that each solution $x_\nu(t)$ has an ω-limit point x_ν^∞ in N_ν, $\nu \in \mathbb{N}_0$. The sequence of ω-limit points x_ν^∞, on the other hand, is bounded and therefore has a convergent subsequence $x_{\nu_\mu}^\infty$ whose limit x_∞^∞, say, lies on M:

$$x_\infty^\infty := \lim_{\mu \to \infty} x_{\nu_\mu}^\infty \in M. \tag{3.5}$$

Near this point x_∞^∞ we can now carry out a local analysis. The relevance of x_∞^∞ is that in any neighborhood of this point there are ω-limit points of solutions of (3.1).

As in the previous section we transform equation (3.2) near x_∞^∞ into a system of the form

$$\begin{aligned} \dot{u} &= A^-(t)u + A_1(t,u,v,w)u + B_1(t,u,v,w)v \\ \dot{v} &= A^+(t)v \qquad\qquad\qquad\quad + B_2(t,u,v,w)v \\ \dot{w} &= \qquad\qquad\quad A_3(t,u,v,w)u + B_3(t,u,v,w)v \end{aligned} \tag{3.6}$$

where the principal fundamental matrices $\Phi^-(t,s)$, $\Phi^+(t,s)$ of $\dot{u} = A^-(t)u$, $\dot{v} = A^+(t)v$, respectively, satisfy estimates of the form

$$\|\Phi^-(t,s)\| \leq \tilde{\gamma}\, e^{-\alpha(t-s)} \quad \text{for all } t \geq s \geq 0,$$

$$\|\Phi^+(t,s)\| \leq \tilde{\gamma}\, e^{-\beta(s-t)} \quad \text{for all } s \geq t \geq 0.$$

The matrices A_i, $i = 1,3$ and B_j, $j = 1,2,3$ are continuous functions which tend to zero uniformly in t as $(u,v,w) \to (0,0,0)$. This property allows us to choose a positive σ such that

$$\begin{aligned} &\|A_i(t,u,v,w)\| \leq \frac{\min\{\alpha,\beta\}}{2\gamma}, \quad i = 1,3, \\ &\|B_j(t,u,v,w)\| \leq \frac{\min\{\alpha,\beta\}}{2\gamma}, \quad j = 1,2,3 \\ &\text{on } \{(t,u,v,w) : t \geq 0,\ \|(u,v,w)\| \leq \sigma\}. \end{aligned} \tag{3.7}$$

The relevance of σ becomes apparent from the following lemma.

Lemma 3.1: If a solution $(u_o(t), v_o(t), w_o(t))$ of (3.6) remains in the ball B_σ of radius σ around $(0,0,0)$ for all $t \geq t_o$ for some t_o, then there exists a point $(0,0,w^*) \in B_\sigma$ such that

$$\lim_{t \to \infty} (u_o(t), v_o(t), w_o(t)) = (0,0,w^*).$$

Proof of Lemma 3.1: The function $(u_o(t), v_o(t), w_o(t))$ may be considered on $[t_o,\infty)$ as a solution of the linear system

$$\begin{aligned} \dot{u} &= [A^-(t) + C_1^o(t)]u + D_1^o(t)v \\ \dot{v} &= [A^+(t) + D_2^o(t)]V \\ \dot{w} &= C_3^o(t)u + D_3^o(t)v \end{aligned} \tag{3.8}$$

where the matrices

$$C_i^o(t) := A_i(t,u_o(t),v_o(t),w_o(t)),\ i = 1,3,$$
$$D_j^o(t) := B_j(t,u_o(t),v_o(t),w_o(t)),\ j = 1,2,3, \tag{3.9}$$

satisfy the estimates

$$\| C_i^o(t) \| \leq \frac{\min\{\alpha,\beta\}}{2\gamma} \text{ and}$$
$$\| D_j^o(t) \| \leq \frac{\min\{\alpha,\beta\}}{2\gamma} \text{ on } [t_o,\infty) \tag{3.10}$$

because of (3.7). We prove the lemma by separately investigating the three components of the solution $(u_o(t),v_o(t),w_o(t))$.

Because of Lemma B.1, Appendix B, the principal fundamental matrix $\hat{\Phi}(t,s)$ of $\dot{v} = [A^+(t) + D_2^o(t)]v$ satisfies the estimate

$$\| \hat{\Phi}(t,s) \| \leq \tilde{\gamma}\, e^{-\frac{\alpha}{2}(s-t)} \quad \text{for all } s \geq t \geq 0.$$

The identity $\hat{\Phi}(t_o,t)v_o(t) = v_o(t_o)$ implies then the estimate

$$\| v_o(t_o) \| \leq \| v_o(t) \| \tilde{\gamma}\, e^{-\frac{\alpha}{2}(t-t_o)} \quad \text{for all } t \geq t_o,$$

thus either $\| v_o(t) \|$ is unbounded as $t \to \infty$ or $v_o(t_o)$ equals zero. The assumptions of the lemma, however, imply that $v_o(t)$ is bounded, hence $v_o(t_o)$ is zero and therefore

$$v_o(t) = 0 \text{ for all } t \geq t_o. \tag{3.11}$$

The function $u_o(t)$ is then a solution of $\dot{u} = [A^-(t) + C_1^o(t)]u$ and the same argument as before shows that

$$\| u_o(t) \| \leq \| u_o(t_o) \| \tilde{\gamma}\, e^{-\frac{\beta}{2}(t-t_o)} \quad \text{for all } t \geq t_o. \tag{3.12}$$

Finally, $w_o(t)$ is a solution of

$$\dot{w} = C_3^o(t)u_o(t). \tag{3.13}$$

This equation turns out to be asymptotically equivalent to

$$\dot{w} = 0, \tag{3.14}$$

i.e. for each solution $w(t)$ of (3.13) there exists a solution $\tilde{w}(t)$ of (3.14) (and vice versa) such that

$$\lim_{t\to\infty} [w(t) - \tilde{w}(t)] = 0.$$

That this equivalence is true can be seen as follows. Because of the boundedness of $C_3^o(t)$ and the estimate (3.12) for $u_o(t)$ the integral

$$\int_{t_o}^{\infty} C_3^o(\tau)u_o(\tau)\, d\tau$$

exists. Thus, the function

$$w_1(t) := -\int_t^{\infty} C_3^o(\tau)u_o(\tau)\, d\tau, \quad t \geq t_o$$

is a solution of (3.13) with

$$\lim_{t\to\infty} w_1(t) = 0.$$

Since the equation

$$w(t) = w_1(t) + \tilde{w}(t)$$

relates the solutions $w(t)$ and $\tilde{w}(t)$ of (3.13) and (3.14), respectively, the asymptotic equivalence of those two equations is obvious. Hence, for the solution $w_o(t)$ of (3.13) there exists a (constant) solution $\tilde{w}(t) = w^*$ of (3.14) such that

$$\lim_{t \to \infty} w_o(t) = w^*.$$

Together with (3.11) and (3.12) this proves the lemma. ∎

The current state of the proof of Theorem 3.1 is as follows. There exists a ball $B_{\bar{\sigma}}$ around $(0,0,0)$, $\bar{\sigma} \leq \sigma$, with the following property: for any positive integer ν there exists a function $(\tilde{u}_\nu(t), \tilde{v}_\nu(t), \tilde{w}_\nu(t))$ which has an ω - limit point in the ball $B_{1/\nu}$ around $(0,0,0)$. Furthermore, from (3.3) we see that each of these functions is a solution of (3.6) in $B_{\bar{\sigma}}$ satisfying the estimates

$$\| \tilde{u}_\nu(t) \| \leq \frac{1}{\nu}, \quad \| \tilde{v}_\nu(t) \| \leq \frac{1}{\nu} \text{ as long as } (\tilde{u}_\nu(t), \tilde{v}_\nu(t), \tilde{w}_\nu(t)) \in B_{\bar{\sigma}}.$$

Let us be more explicit. We may choose a positive ε,

$$\varepsilon < \frac{\bar{\sigma}}{6}, \tag{3.15}$$

with the following property: there exists a function $(u_\varepsilon(t), v_\varepsilon(t), w_\varepsilon(t))$ and a sequence t_ν with $t_\nu \to \infty$ as $\nu \to \infty$ such that the limit $\lim_{\nu \to \infty} (u_\varepsilon(t_\nu), v_\varepsilon(t_\nu), w_\varepsilon(t_\nu))$ exists and satisfies the estimate

$$\| \lim_{\nu \to \infty} (u_\varepsilon(t_\nu), v_\varepsilon(t_\nu), w_\varepsilon(t_\nu)) \| \leq \varepsilon. \tag{3.16}$$

Furthermore, there exists a sequence of intervals J_ν with $t_\nu \in J_\nu$ such that $(u_\varepsilon(t), v_\varepsilon(t), w_\varepsilon(t))$ is a solution of (3.6) on each J_ν, $\nu \in \mathbb{N}_o$.

In addition (3.3) says, in the (u,v,w) - coordinate setting, that

$$\| u_\varepsilon(t) \| \leq \varepsilon, \; \| v_\varepsilon(t) \| \leq \varepsilon \text{ for all } t \in J_\nu, \; \nu \in \mathbb{N}_0. \tag{3.17}$$

From (3.4) we may infer that

$$\begin{aligned} &(u_\varepsilon(t), v_\varepsilon(t), w_\varepsilon(t)) \text{ does not converge} \\ &\text{to any point of the form } (0,0,w). \end{aligned} \tag{3.18}$$

We now distinguish two cases.

Case 1: $(u_\varepsilon(t), v_\varepsilon(t), w_\varepsilon(t))$ remains in $B_{\bar\sigma}$ on an interval of the form $[T,\infty)$. Then Lemma 3.1 says (note that $B_{\bar\sigma} \subset B_\sigma$) that $(u_\varepsilon(t), v_\varepsilon(t), w_\varepsilon(t))$ converges to a point $(0,0,w^*)$ as $t \to \infty$. This contradicts (3.18).

Case 2: $(u_\varepsilon(t), v_\varepsilon(t), w_\varepsilon(t))$ leaves $B_{\bar\sigma}$ again and again. This means that there exists a sequence of instants $T_\nu \in J_\nu$ with the property that

$$\| (u_\varepsilon(t), v_\varepsilon(t), w_\varepsilon(t)) \| < \bar\sigma \text{ on } [t_\nu, T_\nu) \tag{3.19}$$

and

$$\| (u_\varepsilon(T_\nu), v_\varepsilon(T_\nu), w_\varepsilon(T_\nu)) \| = \bar\sigma \text{ for all } \nu \in \mathbb{N}_0. \tag{3.20}$$

Because of (3.17) we have the estimates

$$\| u_\varepsilon(T_\nu) \| \leq \varepsilon, \; \| v_\varepsilon(T_\nu) \| \leq \varepsilon, \; \nu \in \mathbb{N}_0 \tag{3.21}$$

and from this with (3.20) (using a suitable norm)

$$\| w_\varepsilon(T_\nu) \| \geq \bar\sigma - 2\varepsilon \text{ for all } \nu \in \mathbb{N}_0. \tag{3.22}$$

As in the proof of Lemma 3.1 the function $(u_\varepsilon(t),v_\varepsilon(t),w_\varepsilon(t))$ may be considered as a solution on each interval $[t_\nu,T_\nu]$ of the linear equation

$$\begin{aligned} \dot{u} &= [A^-(t) + C_1^\varepsilon(t)]u + D_1^\varepsilon(t)v \\ \dot{v} &= [A^+(t) + D_2^\varepsilon(t)]v \\ \dot{w} &= C_3^\varepsilon(t)u + D_3^\varepsilon(t)v \end{aligned}$$

where, because of (3.7) and (3.19) (note that $\bar{\sigma} \leqq \sigma$), the matrices

$$\begin{aligned} C_i^\varepsilon(t) &:= A_i(t,u_\varepsilon(t),v_\varepsilon(t),w_\varepsilon(t)),\ i = 1,3, \\ D_j^\varepsilon(t) &:= B_j(t,u_\varepsilon(t),v_\varepsilon(t),w_\varepsilon(t)),\ j = 1,2,3, \end{aligned}$$

satisfy the estimates

$$\begin{aligned} \| C_i^\varepsilon(t) \| &\leq \frac{\min\{\alpha,\beta\}}{2\gamma} \quad \text{and} \\ \| D_j^\varepsilon(t) \| &\leqq \frac{\min\{\alpha,\beta\}}{2\gamma} \quad \text{on } [t_\nu,T_\nu],\ \nu \in \mathbb{N}_0. \end{aligned}$$

Hence, we may apply Lemma B.2, Appendix B, to get the inequalities

$$\| w_\varepsilon(T_\nu) \| \leqq \| w_\varepsilon(t_\nu) \| + \| u_\varepsilon(t_\nu) \| + \frac{3}{2} \| v_\varepsilon(T_\nu) \| \quad \text{for all } \nu \in \mathbb{N}_0.$$

With (3.16) and (3.21) this leads to the estimate

$$\| w_\varepsilon(T_\nu) \| \leq \varepsilon + \varepsilon + \frac{3}{2}\varepsilon < 4\varepsilon, \text{ for } \nu \text{ sufficiently large,}$$

which, with the choice (3.15) of ε, can be continued to

$$\| w_\varepsilon(T_\nu) \| < \bar{\sigma} - 2\varepsilon, \text{ for } \nu \text{ sufficiently large.}$$

This, however, contradicts (3.22) and proves the theorem. ∎

4. Saddle Point Property

In this section we consider again the situation where a solution $x(t)$ of a differential equation stays for all future time sufficiently close to a manifold M of stationary solutions. In the previous section we have dealt with the question whether $x(t)$ eventually picks some point on M as its limit. In this section we focus on another aspect of this situation. In the general hyperbolic case to each point $\bar{x}$ on M there is associated a stable manifold $\bar{S}$ on which any solution decays to $\bar{x}$ at an exponential rate. The question arises if there is another way of approaching $\bar{x}$ rather than through the stable manifold $\bar{S}$. In case M is a single point x_o this question is closely related to the so - called saddle point property of x_o. Let us be somewhat more explicit.

A stationary solution x_o of an autonomous differential system

$$\dot{x} = f(x) \tag{4.1}$$

is said to have the *saddle point property* if there exists a neighborhood N of x_o such that any solution of (4.1) having a positive semitrajectory in N lies on the stable manifold of system (4.1) associated with x_o. It is well known (see e.g. Hale [21, III.6]) that x_o has this property if the Jacobian $f_x(x_o)$ has no purely imaginary eigenvalues. Obviously, a nonisolated equilibrium cannot have the saddle point property; in particular, any point on a manifold M of equilibria fails to have this property. The manifold M as a whole, however, may have sort of a saddle point property. In fact, a simple special case of the theorem we are going to prove in this section is the following:

Proposition 4.1: Let $f:\mathbb{R}^n \to \mathbb{R}^n$ be three times continuously differentiable and suppose that (4.1) has a compact C^1-manifold M of stationary solutions which is normally hyperbolic, i.e. for each $\bar{x} \in M$ the Jacobian $f_x(\bar{x})$ has $n - \dim M$ eigenvalues with real parts different from zero.
Then there exists an $\mathbb{R}^n$-neighborhood N of M such that any solution of (4.1) having a positive semitrajectory in N lies on the stable manifold of some stationary solution on M.

The class of equations we are going to consider in this section is the same as that in the previous one. The assumptions (i) through (iii) of Section 3 guarantee that each stationary solution $\bar{x} \in M$ is associated with a stable manifold which is an integral manifold of dimension $1 + \operatorname{rank} P^-$. It consists of all integral curves which approach $\bar{x}$ at an exponential rate which is at least α. With this we are prepared for the formulation of the main result of this section.

Theorem 4.1: Under the hypotheses (i), (ii), (iii) of Section 3 there exists an $\mathbb{R}^n$-neighborhood N of M with the following property: if an integral curve $(t,x(t))$ of system (3.1) eventually remains in $\mathbb{R}^+ \times N$, then it lies on the stable manifold of some stationary solution of (3.1) on M.

Proof: From Theorem 3.2 we know that there exists a point $x^* \in M$ such that

$$\lim_{t\to\infty} x(t) = x^*. \tag{4.2}$$

That this already implies the validity of Theorem 4.1 can be seen as follows. Without loss of generality we may use local coordinates so that our equation near x^* is represented as in (3.6), i.e.

$$\begin{aligned} \dot{u} &= A^-(t)u + r_1(t,u,v,w) \\ \dot{v} &= A^+(t)v + r_2(t,u,v,w) \\ \dot{w} &= \qquad\qquad r_3(t,u,v,w) \end{aligned} \tag{4.3}$$

with $r_1(t,0,0,w) \equiv 0$, $r_2(t,0,0,w) \equiv 0$, $r_3(t,0,0,w) \equiv 0$, where the coordinate origin $(u,v,w) = (0,0,0)$ corresponds to x^*. Let $(u(t),v(t),w(t))$ denote the solution associated with $x(t)$. A consequence of (4.2) is then that $(u(t),v(t),w(t))$ eventually remains in an arbitrarily small neighborhood of $(0,0,0)$ (not only in a neighborhood of the linear manifold $u=0$, $v=0$ of stationary solutions). By well known properties of integral manifolds (see e.g. Hale [21, Chapter VII], Knobloch and Kappel [30, Kapitel V]) this means that the integral curve $(t,u(t), v(t),w(t))$ lies on any local center - stable manifold of (4.3) through the coordinate origin, i.e. an integral manifold with representation $v=r(t,u,w)$. On the other hand, in our situation there exists a family of stable manifolds

$$v = s_1(t,u,\lambda), \quad w = s_2(t,u,\lambda) \tag{4.4}$$

corresponding to the family $(0,0,\lambda)$ of stationary solutions of (4.3) near $(0,0,0)$. Because of the identity $s_2(t,0,\lambda) \equiv \lambda$ we get

$$\det\, (s_2)_\lambda(t,0,0) \neq 0 \text{ for each } t \geq 0$$

and this shows that (4.4) is a parameter representation of an integral manifold of the form

$$v = s(t,u,w).$$

Thus, (4.4) describes a center - stable manifold of (4.3) near (0,0,0). With the above argument we see then that the integral curve $(t,u(t), v(t),w(t))$ must lie on one member of the family (4.4) of stable manifolds. In fact, it lies on the stable manifold through (0,0,0) since otherwise the solution $(u(t),v(t),w(t))$ would converge to a neighboring stationary solution of (0,0,0). This proves the theorem. ■

5. Equations with a Nonstationary Integral Manifold

In the two previous sections our fundamental Theorem 2.1 was modified so that the ω - limit set of the given solution $x(t)$ was not required to lie on the manifold M of stationary solutions. In this section we return to the assumption of Section 2 as far as the ω - limit set of $x(t)$ is concerned; however, we do not suppose any more that the integral manifold $\mathbb{R}^+ \times M$ consists of stationary solutions. This assumption is replaced by a weaker assumption concerning the flow on $\mathbb{R}^+ \times M$. In order to avoid some technicalities we suppose a priori that the underlying differential equation is linearly decoupled and that the integral manifold $\mathbb{R}^+ \times M$ has the form $\{(t,0,0,w): t \geq 0, \|w\| \text{ small}\}$. To be more explicit, we consider systems of the form

$$\begin{aligned} \dot{u} &= A^-(t)u + r_1(t,u,v,w) \\ \dot{v} &= A^+(t)v + r_2(t,u,v,w) \\ \dot{w} &= A^o(t)w + r_3(t,u,v,w), \end{aligned} \tag{5.1}$$

$u \in \mathbb{R}^{n_-}$, $v \in \mathbb{R}^{n_+}$, $w \in \mathbb{R}^{n_o}$, $n = n_- + n_+ + n_o$, satisfying the following hypotheses:

(i) the nonlinearities $r_i(t,u,v,w)$, $i = 1,2,3$, are continuous on $\mathbb{R}^+ \times \mathbb{R}^n$ and of class C^3 as functions of u, v, w,

(ii) the functions $r_i(t,u,v,w)$, $i = 1,2,3$, and their first order derivatives with respect to u, v, w tend to zero uniformly in t as $\|(u,v,w)\| \to 0$,

(iii) $r_1(t,0,0,w) = 0$, $r_2(t,0,0,w) = 0$ for all $t \geq 0$ and $\|w\| < r_o$ for some positive r_o,

(iv) the matrices $A^-(t)$, $A^+(t)$, $A^o(t)$ are continuous and bounded on $\mathbb{R}^+$,

(v) there exist positive constants $\alpha, \beta, \gamma, \delta$, $\delta < \beta$, such that the principal fundamental matrices $\Phi^-(t,s)$, $\Phi^+(t,s)$, $\Phi^o(t,s)$ of $\dot{u} = A^-(t)u$, $\dot{v} = A^+(t)v$, $\dot{w} = A^o(t)w$, respectively, satisfy the estimates

$$\|\Phi^-(t,s)\| \leq \gamma e^{-\alpha(t-s)} \quad \text{for all } t \geq s \geq 0,$$

$$\|\Phi^+(t,s)\| \leq \gamma e^{-\beta(s-t)} \quad \text{for all } s \geq t \geq 0,$$

$$\|\Phi^o(t,s)\| \leq \gamma e^{\delta \cdot (t-s)} \quad \text{for all } t \geq s \geq 0,$$

(vi) the zero solution of

$$\dot{w} = A^o(t)w + r_3(t,0,0,w) \tag{5.2}$$

is uniformly stable,

(vii) there exists a positive constant ε such that for each solution $w(t)$ of (5.2) the principal fundamental matrix $\Phi(t,s)$ of

$$\dot{w} = [A^o(t) + (r_3)_w(t,0,0,w(t))]w$$

satisfies the estimate

$$\|\Phi(t,s)\| \leq \varepsilon \quad \text{for all } s \geq t \geq 0.$$

For system (5.1), subject to those conditions, we get the following result.

Theorem 5.1: Consider system (5.1) having a nonstationary integral manifold satisfying the conditions (i) through (vii) above. Let $(u(t),v(t),w(t))$ be any solution with ω-limit set Ω. Suppose that

(I) $0 \in \Omega$,

(II) there exists an $\mathbb{R}^n$-neighborhood U of $(u,v,w) = (0,0,0)$ such that $\Omega \cap U \subset \{(0,0,w): w \in \mathbb{R}^{n_o}\}$.

Then $\lim_{t \to \infty} (u(t),v(t),w(t)) = (0,0,0)$.

Proof: Although system (5.1) is more general than system (2.11) the statement of Lemma 2.1 remains valid also for (5.1). The verification of this resembles the proof of Lemma 2.1 and is therefore only sketched. The hypotheses (i) and (ii) of Section 5 allow us to replace system (5.1) by a system of the form

$$\begin{aligned} \dot{\bar{z}} &= A(t)\bar{z} + \bar{r}(t,\bar{z},\bar{v}) \\ \dot{\bar{v}} &= A^{+}(t)\bar{v} + \bar{r}_3(t,\bar{z},\bar{v}), \end{aligned} \tag{5.3}$$

$\bar{z} = (u,w)$, $\bar{v} = v$, $A(t) := \operatorname{diag}(A^{-}(t),A^{o}(t))$, which coincides with system (5.1) on a set $\{(t,\bar{z},\bar{v}): t \geq 0, \ \|(\bar{z},\bar{v})\| < r\}$ for some positive r and moreover satisfies the global boundedness conditions of Theorem 2.2. The hypotheses (iv) and (v) guarantee that the principal fundamental matrix $\bar{\Phi}(t,s)$ of $\dot{\bar{z}} = A(t)\bar{z}$ satisfies for some positive $\bar{\gamma}$ the estimate

$$\|\bar{\Phi}(t,s)\| \leq \bar{\gamma} e^{\delta(t-s)} \quad \text{for all } t \geq s \geq 0. \tag{5.4}$$

Since δ is by assumption smaller than β we can verify, by suitable choice of r, the dichotomy condition (3) of Theorem 2.2. The coupling constant ρ in condition (4) of Theorem 2.2 may be any number between δ and β. Thus, Theorem 2.2 applies to system (5.3) and yields an integral manifold $\bar{v} = s(t,\bar{u},\bar{w})$. In order to get the validity of (ii) of Lemma 2.1 we notice that the proof of Lemma 2.1 remains unchanged up to the relation (2.19) which now reads

$$\| \bar{z}(t,t_o,\bar{z}_o) \| \leqq \gamma e^{\delta(t-t_o)} \quad \text{for all } t \geqq t_o .$$

This is because (2.15) has to be replaced by (5.4). This change implies that instead of (2.20) we get

$$\| S(t_o)\bar{z}_o \| \leqq \| S(t) \| \bar{\gamma} \| \bar{z}_o \| \gamma e^{(\delta-\beta)(t-t_o)} \quad \text{for all } t \geqq t_o .$$

Since $\delta - \beta$ is negative we arrive at the same contradiction as in the proof of Lemma 2.1 which says then that $s_{(\bar{u},\bar{w})}(t,0,0)$ vanishes for all $t \geqq 0$. The identity $s(t,0,\bar{w}) \equiv 0$ finally follows from the maximality of s and hypotheses (iii) and (vi) of this section. Thus we have shown that the statement of Lemma 2.1 holds true also for system (5.1) and we have the function s at our disposal.

As in the proof of Theorem 2.1 we subject system (5.1) to the transformation

$$\begin{aligned} \hat{u} &= u \\ \hat{v} &= v - s(t,u,w) \\ \hat{w} &= w . \end{aligned}$$

Rather than (2.22), this time we get a system of the form

$$\begin{aligned}
\dot{\hat{u}} &= A^-(t)\hat{u} + A_1(t,\hat{u},\hat{v},\hat{w})\hat{u} + B_1(t,\hat{u},\hat{v},\hat{w})\hat{v} \\
\dot{\hat{v}} &= A^+(t)\hat{v} \qquad\qquad\qquad\qquad + B_2(t,\hat{u},\hat{v},\hat{w})\hat{v} \\
\dot{\hat{w}} &= A^o(t)\hat{w} + r_3(t,0,0,\hat{w}) + A_3(t,\hat{u},\hat{v},\hat{w})\hat{u} + B_3(t,\hat{u},\hat{v},\hat{w})\hat{v},
\end{aligned} \tag{5.5}$$

where the matrices A_i, B_j are continuous functions of all their variables which tend to zero uniformly in $t \in \mathbb{R}^+$ as $\|(\hat{u},\hat{v},\hat{w})\| \to 0$. For the moment, we continue as in the proof of Theorem 2.1. By assumption there exists a solution $(\hat{u}(t),\hat{v}(t),\hat{w}(t))$ such that

$$(2.23),(2.24),(2.25),(2.26) \text{ are valid.} \tag{5.6}$$

Adopting also the notation from Section 2 we may choose $\hat{\sigma}$ so small that

$$\|A_i(t,\hat{u}(t),\hat{v}(t),\hat{w}(t))\| \le \frac{1}{2\gamma\varepsilon}\min\{\alpha,\beta\},\quad i=1,3,$$
$$\|B_j(t,\hat{u}(t),\hat{v}(t),\hat{w}(t))\| \le \frac{1}{2\gamma\varepsilon}\min\{\alpha,\beta\},\quad j=1,2,3,$$

on each interval $[t_\nu,T_\nu]$, $\nu \in \mathbb{N}_0$. Consequently, for each $\nu \in \mathbb{N}_0$ the function $(\hat{u}(t),\hat{v}(t),\hat{w}(t))$ is a solution of the equation

$$\begin{aligned}
\dot{\hat{u}} &= A^-(t)\hat{u} + C_1(t)\hat{u} + D_1(t)\hat{v} \\
\dot{\hat{v}} &= A^+(t)\hat{v} \qquad\qquad + D_2(t)\hat{v} \\
\dot{\hat{w}} &= A^o(t)\hat{w} + r_3(t,0,0,\hat{w}) + C_3(t)\hat{u} + D_3(t)\hat{v}
\end{aligned} \tag{5.7}$$

on $[t_\nu,T_\nu]$ where the matrices

$$C_i(t) := A_i(t,\hat{u}(t),\hat{v}(t),\hat{w}(t)),\quad i=1,3,$$
$$D_j(t) := B_j(t,\hat{u}(t),\hat{v}(t),\hat{w}(t)),\quad j=1,2,3,$$

have upper bound $\frac{1}{2\gamma\varepsilon}\min\{\alpha,\beta\}$. Since (5.7) is a nonlinear system whereas the corresponding system (2.22) is linear, here we have to depart from the proof of Theorem 2.1.

Let us consider the system

$$\dot{\hat{w}} = A^{o}(t)\hat{w} + R(t,\hat{w}) \tag{5.8}$$

where $R(t,\hat{w})$ coincides with $r_3(t,0,0,\hat{w})$ for $t \geq 0$ and $\|\hat{w}\| \leq R_o$ and vanishes for $t \geq 0$ and $\|\hat{w}\| \geq 2R_o$ for some positive R_o which is still to be determined suitably. We denote by $W(t,\tilde{t},\tilde{w})$ the solution of (5.8) with $W(\tilde{t},\tilde{t},\tilde{w}) = \tilde{w}$. From the stability assumption (vi) we get

$$\lim_{\tilde{w}\to 0} W(t,\tilde{t},\tilde{w}) = 0 \text{ uniformly in } t,\ \tilde{t} \text{ with } t \geq \tilde{t} \geq 0. \tag{5.9}$$

Furthermore, for $\|\tilde{w}\|$ suitably small, $W_{\tilde{w}}(t,\tilde{t},\tilde{w})$ is the principal fundamental matrix of

$$\dot{\hat{w}} = [A^{o}(t) + (r_3)_w(t,0,0,W(t,\tilde{t},\tilde{w}))]\hat{w}$$

and this implies the identity

$$(W_{\tilde{w}}(t,\tilde{t},\tilde{w}))^{-1} = W_{\tilde{w}}(\tilde{t},t,\tilde{w})$$

and furthermore, with assumption (vii), the estimate

$$\|(W_{\tilde{w}}(t,\tilde{t},\tilde{w}))^{-1}\| \leq \varepsilon \text{ for all } t \geq \tilde{t} \geq 0. \tag{5.10}$$

After these preparations we are able to convert the nonlinear equation (5.7) into a linear one which is amenable to our Lemma B.2, Appendix B. Near $(\hat{u},\hat{v},\hat{w}) = (0,0,0)$ we make, for each $\nu \in \mathbb{N}_o$, the change of coordinates

$$\hat{u} = \tilde{u}$$
$$\hat{v} = \tilde{v}$$
$$\hat{w} = W(t,t_\nu,\tilde{w})$$

which gives the new system in the form

$$\begin{aligned}\dot{\tilde{u}} &= [A^-(t) + C_1(t)]\tilde{u} + D_1(t)\tilde{v} \\ \dot{\tilde{v}} &= [A^+(t) + D_2(t)]\tilde{v} \\ \dot{\tilde{w}} &= (W_{\tilde{w}}(t,t_\nu,\tilde{w}))^{-1}C_3(t)\tilde{u} + (W_{\tilde{w}}(t,t_\nu,\tilde{w}))^{-1}D_3(t)\tilde{v}.\end{aligned} \tag{5.11}$$

On each interval $[t_\nu,T_\nu]$ the function

$$(\tilde{u}_\nu(t),\tilde{v}_\nu(t),\tilde{w}_\nu(t)) := (\hat{u}(t),\hat{v}(t),W(t_\nu,t,\hat{w}(t))) \tag{5.12}$$

is a solution of a linear system of the form

$$\begin{aligned}\dot{\tilde{u}} &= [A^-(t) + C_1(t)]\tilde{u} + D_1(t)\tilde{v} \\ \dot{\tilde{v}} &= [A^+(t) + D_2(t)]\tilde{v} \\ \dot{\tilde{w}} &= \tilde{C}_3(t)\tilde{u} + \tilde{D}_3(t)\tilde{v}\end{aligned} \tag{5.13}$$

where, because of (5.10), the matrices $\tilde{C}_3(t)$ and $\tilde{D}_3(t)$ are bounded above by $\frac{1}{2\gamma}\min\{\alpha,\beta\}$. Since we may assume without loss of generality that $\varepsilon \geq 1$ also the matrices $C_1(t)$, $D_1(t)$, $D_2(t)$ have this upper bound on each interval $[t_\nu,T_\nu]$ and so we may apply Lemma B.2, Appendix B, to equation (5.13). This provides for each $\nu \in \mathbb{N}_0$ the identity

$$\|\tilde{w}_\nu(T_\nu)\| \leq \|\tilde{w}_\nu(t_\nu)\| + \|\tilde{u}_\nu(t_\nu)\| + \frac{3}{2}\|\tilde{v}_\nu(T_\nu)\| \tag{5.14}$$

which is (2.29) except for the ν - dependence of the functions $\tilde{u}$, $\tilde{v}$ and $\tilde{w}$. The sequence $(\tilde{u}_\nu(T_\nu),\tilde{v}_\nu(T_\nu),\tilde{w}_\nu(T_\nu))$ is bounded and therefore it has a convergent subsequence $(\tilde{u}_{\nu_\mu}(T_{\nu_\mu}),\tilde{v}_{\nu_\mu}(T_{\nu_\mu}),\tilde{w}_{\nu_\mu}(T_{\nu_\mu}))$ with

$$\lim_{\mu\to\infty} \tilde{u}_{\nu_\mu}(T_{\nu_\mu}) = 0, \tag{5.15}$$

$$\lim_{\mu\to\infty} \tilde{v}_{\nu_\mu}(T_{\nu_\mu}) = 0 \tag{5.16}$$

(see (5.6), (2.24) and notice that $\tilde{u}_\nu(t) \equiv \hat{u}(t)$, $\tilde{v}_\nu(t) \equiv \hat{v}(t)$ for all $\nu \in \mathbb{N}_o$). Furthermore, using the identity

$$\hat{w}(t) \equiv W(t,t_\nu,\tilde{w}_\nu(t))$$

(see (5.12)) we get the limiting relations

$$\lim_{\mu \to \infty} \tilde{w}_{\nu_\mu}(t_{\nu_\mu}) = 0 \tag{5.17}$$

(see (5.6), (2.23) and notice that $\hat{w}(t_\nu) = \tilde{w}_\nu(t_\nu)$ for all $\nu \in \mathbb{N}_o$) and

$$\lim_{\mu \to \infty} \tilde{w}_{\nu_\mu}(T_{\nu_\mu}) \neq 0. \tag{5.18}$$

The last limit is different from zero since otherwise one would get with (5.9) the relation $\lim_{\mu \to \infty} \hat{w}(T_{\nu_\mu}) = 0$ which, together with (5.15), (5.16) contradicts (5.6), (2.26). A final relation we need for the completion of our proof is

$$\lim_{\mu \to \infty} \tilde{u}_{\nu_\mu}(t_{\nu_\mu}) = 0. \tag{5.19}$$

It follows readily from (5.6), (2.23). As in the proof of Theorem 2.1 a contradiction is obtained from the noncompatibility of the relations (5.14), (5.16), (5.17), (5.18) and (5.19). This completes the proof of Theorem 5.1. ∎

6. The Basic Selection Model from Population Genetics for Overlapping Generations

As an application of Theorem 2.3 we consider the well known Fisher-Wright-Haldane model from population genetics, to wit, we consider the single (autosomal) locus model with n alleles, the only influence on the population being natural selection. On a plausibility level we briefly sketch the derivation of the continuous time model which corresponds to the case of overlapping generations. For more details and the biological background we refer to Crow and Kimura [13], Edwards [14] and Hadeler [19],[20].

Consider a population of diploid individuals (i.e. each individual has two parents, mother and father) and fix a certain feature F of the individuals which is genetically determined such as the blood group, skin color etc. Then divide the total population in subgroups of individuals which are genetically homogeneous with respect to this particular property F. Suppose that there are $n \geq 2$ genes $a_1, \ldots, a_n$ determining F. Those genes are called *alleles* for the feature F. Since the genetical information for diploid organisms is contained in a pair (a_j, a_k) of alleles, one allele from each parent, the population contains n^2 different genotypes (a_j, a_k), $j,k = 1,\ldots,n$. Population genetics is concerned with the evolution of the different genotypes; to be more explicit, the question is how the different genotype proportions develop or, in other words, how the sizes of the subgroups G_{jk} develop relative to each other, where G_{jk} denotes the group of individuals with allele combination (a_j, a_k). For the mathematical description of this problem one introduces the so-called *genotype frequencies* α_{jk} which are the relative frequencies of the (a_j, a_k)-genotypes or, equivalently, the sizes of the respective groups G_{jk}. For biological reasons one may assume that $\alpha_{jk} = \alpha_{kj}$ for $j,k = 1,\ldots,n$. As mathematical properties we have

$$\alpha_{jk} \geq 0, \quad \sum_{j,k=1}^{n} \alpha_{jk} = 1.$$

Since on the genotype level the mathematical model cannot be treated in a satisfactory way one makes the so-called *random mating* assumption which means that one assumes that

$$\alpha_{jk} = x_j x_k, \quad j,k = 1,\dots,n, \tag{6.1}$$

where x_i is the *gene frequency* of the allele a_i, i.e. the relative frequency of a_i in the total population. Under this assumption the evolution of the population can be described completely by the gene frequencies x_i. The following relations are obvious:

$$x_i \geq 0, \quad \sum_{i=1}^{n} x_i = 1, \quad x_i = \sum_{k=1}^{n} \alpha_{ik}, \quad i = 1,\dots,n.$$

The selection in the population is described by the so-called *viability parameters* $f_{jk} = f_{kj}$, $j,k = 1,\dots,n$, which are nonnegative. f_{jk} can be interpreted as a measure for the reproductivity of the genotype (a_j,a_k).

In order to derive the differential equation for the gene frequency vector $x = (x_1,\dots,x_n)^T$ we denote temporarily the absolute frequency of the allele a_j at time t by $N_j(t)$, thus getting

$$x_j(t) = N_j(t) / \sum_{i=1}^{n} N_i(t). \tag{6.2}$$

The production of (a_j,a_k)-genotypes in a time interval of length Δt is then given by

$$f_{jk} N_j(t) N_k(t) \Delta t / \sum_{i=1}^{n} N_i(t).$$

Hence, the total number of newly produced alleles a_j is

$$N_j(t+\Delta t) - N_j(t) = N_j(t) \left[\sum_{k=1}^{n} f_{jk} N_k(t) \right] \Delta t / \sum_{i=1}^{n} N_i(t).$$

Letting Δt tend to zero we get the set of differential equations

$$\dot{N}_j = N_j[\sum_{k=1}^{n} f_{jk}N_k]/\sum_{i=1}^{n} N_i, \quad j=1,\dots,n,$$

and using (6.2) we arrive at the system of differential equations

$$\dot{x}_j = x_j[\sum_{k=1}^{n} f_{jk}x_k - \sum_{r,s=1}^{n} f_{rs}x_r x_s], \quad j=1,\dots,n,$$

for the gene frequencies. Introducing the symmetric matrix $F=(f_{jk})$ and the so-called *mean viability*

$$\Phi(x) := x^T F x \tag{6.3}$$

the last system can be written in vector form

$$\frac{d}{dt}\begin{pmatrix} x_1 \\ \vdots \\ x_n \end{pmatrix} = \begin{pmatrix} x_1 & & 0 \\ & \ddots & \\ 0 & & x_n \end{pmatrix} F \begin{pmatrix} x_1 \\ \vdots \\ x_n \end{pmatrix} - \Phi(x)\begin{pmatrix} x_1 \\ \vdots \\ x_n \end{pmatrix} =: G(x). \tag{6.4}$$

It is well known (see e.g. Hadeler [20, p.148]) that this differential equation leaves the simplex

$$\Lambda^{n-1} := \{x \in \mathbb{R}^n : \sum_{i=1}^{n} x_i = 1,\ x_i \geq 0,\ i=1,\dots,n\} \tag{6.5}$$

as well as each of its subsimplices invariant. For any $x_o \in \Lambda^{n-1}$ we denote by $x(t,x_o)$ the solution of (6.4) with $x(0,x_o) = x_o$ and by $\Omega(x_o)$ the ω-limit set of this solution. The set of stationary solutions of (6.4) on Λ^{n-1} is denoted by E^{n-1},

$$E^{n-1} := \{x \in \Lambda^{n-1} : \operatorname{diag}(x)Fx = \Phi(x)x\}.$$

It is the aim of this section to apply Theorem 2.3 to equation (6.4) and so to obtain information about the asymptotic bahavior of the solutions of this system on the state space Λ^{n-1}. In case $n = 3$ any solution on Λ^{n-1} converges to a stationary solution as time goes to infinity. This has been proved by an der Heiden [25] with methods heavily depending on the fact that the state space of (6.4) in this case is two - dimensional. Thus, the proof does not carry over to higher dimensions. The general case has been treated by Aulbach and Hadeler [7] under the assumption that all viability parameters f_{jk} are positive. Our Theorem 6.2 below extends the results in [7] to the general case of nonnegative f_{jk}.

First information concerning the convergence problem of the solutions of (6.4) can be drawn from the well known Theorem of LaSalle which, adapted to our situation, is stated for completeness.

Theorem 6.1 (LaSalle [32, Chapter 1, Theorem 6.3]): Let K be any subset of Λ^{n-1} and suppose that there exists a function $V \in C^1(\mathbb{R}^n, \mathbb{R})$ such that the function

$$\dot{V}(x) := V_x(x)G(x)$$

is nonnegative on K. Then for any $x_o \in K$ such that $x(t,x_o)$ is in K for all $t \geq 0$ one has

$$\Omega(x_o) \subset \{x \in \bar{K}: \dot{V}(x) = 0\}.$$

In order to prove that the mean viability $\Phi(x)$ is a Ljapunov function satisfying the hypotheses of LaSalle's Theorem we need a first lemma which in similar form can be found in the standard literature (see e.g. Hadeler [20, p.150]).

Lemma 6.1: For all $x \in \Lambda^{n-1}$ the following holds true:

(i) $$x^T F \operatorname{diag}(x) F x \geq \Phi(x)^2,$$

(ii) equality in (i) holds if and only if

$$\operatorname{diag}(x) F x = \Phi(x) x.$$

Proof: With the abbreviations

$$y := \operatorname{diag}(x_1^{1/2}, \ldots, x_n^{1/2}) F x,$$
$$z := (x_1^{1/2}, \ldots, x_n^{1/2})^T$$

we get the relations (note that F is symmetric and that $\Sigma_{i=1}^n x_i = 1$)

$$y^T y = x^T F \operatorname{diag}(x) F x, \quad y^T z = \Phi(x), \quad z^T z = 1. \tag{6.6}$$

Those relations lead to the identity

$$x^T F \operatorname{diag}(x) F x - \Phi(x)^2 = (y^T y)(z^T z) - (y^T z)^2$$

and using the Cauchy Schwarz inequality for y and z we obtain the asserted inequality (i). Moreover, equality in (i) holds if and only if y and z are linearly dependent. This means that there exists a real δ such that $y = \delta z$. Multiplying the equation $y = \delta z$ from the left first with $\operatorname{diag}(x_1^{1/2}, \ldots, x_n^{1/2})$ and then with $e^T = (1, \ldots, 1)$ we get

$$\operatorname{diag}(x) F x = \delta x, \quad \delta = \Phi(x)$$

and the proof is complete. ∎

As a consequence of LaSalle's Theorem we get the next lemma.

Lemma 6.2: For any $x_o \in \Lambda^{n-1}$ the ω - limit set $\Omega(x_o)$ is contained in the set E^{n-1} of stationary solutions of (6.4).

Remark: If the set E^{n-1} consists of finitely many points only, then it follows immediately from this lemma that any solution of (6.4) converges to a stationary solution. This is due to the connectedness of the ω - limit set. In general, however, the set E^{n-1} contains continua of stationary solutions (see Appendix C) and so Lemma 6.2 represents only a first step in the analysis of the convergence problem. On the other hand, just at this state of knowledge our Theorem 2.3 comes into play.

Proof of Lemma 6.2: With $K := \Lambda^{n-1}$ and $V := \Phi$ we apply Theorem 6.1. The function $\dot{\Phi}(x)$ has the representation

$$\dot{\Phi}(x) = 2x^T F[\mathrm{diag}(x)Fx - \Phi(x)x] = 2[x^T F\,\mathrm{diag}(x)Fx - \Phi(x)^2] \qquad (6.7)$$

and therefore it is nonnegative on K due to Lemma 6.1. Because of the invariance of $K = \Lambda^{n-1}$, the solution $x(t,x_o)$ does not leave Λ^{n-1} and so we may apply Theorem 6.1 which yields the inclusion

$$\Omega(x_o) \subset \{x \in \Lambda^{n-1} : \dot{\Phi}(x) = 0\}. \qquad (6.8)$$

In order to complete the proof it only remains to be shown that the set on the right hand side of (6.8) is contained in E^{n-1}.
However, with (6.7) and Lemma 6.1 we immediately get for any $x \in \Lambda^{n-1}$ with $\dot{\Phi}(x) = 0$ the relation $\mathrm{diag}(x)Fx = \Phi(x)x$ and this means that x is a stationary solution of (6.4). ∎

The formulation of the next two lemmas and the main result of this section requires some additional notation. By $N_1,\ldots,N_{2^n}$ we denote (in any order) the 2^n different subsets of

$$N := \{1,\ldots,n\},$$

for definiteness we set $N_{2^n} := \emptyset$. For $j = 1,\ldots,2^n$ let

$$n_j := \operatorname{card} N_j$$

be the cardinal number of N_j, $j = 1,\ldots,n$. Then associated with each $x \in \Lambda^{n-1}$ we define the

$$\textit{index } j = j(x)$$

as the uniquely determined number $j \in \{1,\ldots,2^n\}$ such that

$$x \in R_j := \{x \in \mathbb{R}^n : x_i > 0 \text{ for } i \in N_j,\ x_k = 0 \text{ for } k \in N \setminus N_j\}.$$

For each $j \in \{1,\ldots,2^n-1\}$ we define the submatrix F_j of F which is obtained from F by deleting the rows and columns with indices in $N \setminus N_j$, in technical terms

$$F_j := (f_{i_\nu k_\nu}),\ \nu = 1,\ldots,n_j,$$

where i_ν, k_ν are elements of N_j such that $i_1 < \ldots < i_{n_j}$ and $k_1 < \ldots < k_{n_j}$. Then, for each $x \in \Lambda^{n-1}$ we define the

$$\textit{defect } d = d(x) := \dim \ker F_{j(x)}$$

where $j(x)$ is the index of x. Finally, we call a point $x \in \Lambda^{n-1}$

$$\textit{regular} \text{ if } \sum_{k=1}^{n} f_{ik}x_k \neq \Phi(x) \text{ for all } i \in N \setminus N_{j(x)},$$

otherwise it is called *exceptional*. It should be noted that the set of regular points is open (in the relative topology) in Λ^{n-1} and that all points in the relative interior of Λ^{n-1} are regular since any interior point of Λ^{n-1} has an index j_o such that $N_{j_o} = N$.

The proof of the next lemma follows quite easily from Theorem C.2 in Appendix C.

<u>Lemma 6.3</u>: Let $\bar{x}$ be a regular stationary solution (i.e. a stationary solution which is a regular point) of (6.4) and let $d = d(\bar{x})$ be its defect.
Then there exists an $\mathbb{R}^n$- neighborhood U of $\bar{x}$ such that $E^{n-1} \cap U$ is a smooth manifold of dimension d.

<u>Proof</u>: Rearranging the components of the state vector x we may assume without loss of generality that the index j of $\bar{x}$ is such that the subset N_j of N is composed of the first n_j positive integers. Then the regularity assumption of $\bar{x}$ implies that the relation (C.7) in Appendix C is satisfied and so we may apply Theorem C.2, Appendix C. This proves Lemma 6.3. ∎

In order to be able to apply Theorem 2.3 we provide one more lemma.

<u>Lemma 6.4</u>: Let $\bar{x}$ be a regular stationary solution of (6.4) and let $d = d(\bar{x})$ be its defect.

Then the Jacobian $G_x(\bar{x})$ of the right hand side G of (6.4) at $\bar{x}$ has $n-d$ eigenvalues with real parts different from zero.

Proof: Again we rearrange the components of $\bar{x}$ so that

$$N_{j(\bar{x})} = \{1,\ldots,n_{j(\bar{x})}\}.$$

This means that $\bar{x}$ and F can be partitioned in the form

$$\bar{x} = \begin{pmatrix} \tilde{x} \\ 0 \end{pmatrix}, \quad F = \begin{pmatrix} F_j & F_{12} \\ F_{21} & F_{22} \end{pmatrix}, \tag{6.9}$$

where $\tilde{x} \in \mathbb{R}^{n_j}$, $F_j \in \mathbb{R}^{n_j \times n_j}$. Note that the regularity of $\bar{x}$ in this setting means that

$$\text{none of the components of the vector } F_{21}\tilde{x} \text{ equals } \Phi(\bar{x}). \tag{6.10}$$

In order to compute the Jacobian of G at $\bar{x}$ we turn, for the moment, from equation (6.4) to the equation for the perturbed motion $y = x - \bar{x}$. Thus, we consider the equation

$$\begin{aligned}\dot{y} &= \operatorname{diag}(\bar{x}+y)F(\bar{x}+y) - [(\bar{x}+y)^T F(\bar{x}+y)](\bar{x}+y) = \\ &\operatorname{diag}(\bar{x})Fy + \operatorname{diag}(y)F\bar{x} - 2(\bar{x}^T Fy)\bar{x} - \Phi(\bar{x})y + o(y^T y) = \\ &[\operatorname{diag}(\bar{x})F + \operatorname{diag}(F\bar{x}) - 2\bar{x}\bar{x}^T F - \Phi(\bar{x})I_n]y + o(y^T y),\end{aligned}$$

where $o(y^T y)$ refers to the limit $y \to 0$. Employing the particular representation (6.9) of $\bar{x}$ and F we get $G_x(\bar{x})$ in the form

$$G_x(\bar{x}) = \begin{pmatrix} G_{11}(\bar{x}) & G_{12}(\bar{x}) \\ 0 & G_{22}(\bar{x}) \end{pmatrix} \tag{6.11}$$

where

$$G_{11}(\bar{x}) := \mathrm{diag}(\tilde{x})F_j + \mathrm{diag}(F_j\tilde{x}) - 2\tilde{x}\tilde{x}^T F_j - \Phi(\bar{x})I_{n_j},$$
$$G_{12}(\bar{x}) := [\mathrm{diag}(\tilde{x}) - 2\tilde{x}\tilde{x}^T]F_{12},$$
$$G_{22}(\bar{x}) := \mathrm{diag}(F_{21}\tilde{x}) - \Phi(\bar{x})I_{n-n_j}. \tag{6.12}$$

Since $\bar{x}$ is a stationary solution of (6.4) and since the components of $\tilde{x}$ are positive, $\tilde{x}$ satisfies the relation

$$F_j\tilde{x} = \Phi(\bar{x})e^j \tag{6.13}$$

where e^j denotes the n_j-dimensional column vector $(1,\ldots,1)^T$. Thus, $G_{11}(\bar{x})$ reduces to

$$G_{11}(\bar{x}) = \mathrm{diag}(\tilde{x})F_j - 2\Phi(\bar{x})\tilde{x}(e^j)^T. \tag{6.14}$$

The regularity of $\bar{x}$ (see (6.10)) implies that the lower right block $G_{22}(\bar{x})$ of $G_x(\bar{x})$ is a real nonsingular diagonal matrix and so the claim of the lemma reduces to the

claim: $G_{11}(\bar{x})$ has $n_j - d$ eigenvalues with real parts $\neq 0$. (6.15)

In order to prove (6.15) we distinguish two cases, treating the trivial one first.

Case 1: $\Phi(\bar{x}) = 0$. This implies that $F_j = O_{n_j \times n_j}$, hence $d = n_j$ and $G_{11}(\bar{x}) = O_{n_j \times n_j}$. This means that (6.15) is valid with $n_j - d = 0$.

Case 2: $\Phi(\bar{x}) > 0$, i.e. $F_j \neq O_{n_j \times n_j}$. Introducing the abbreviation

$$\tilde{X} := \mathrm{diag}((\tilde{x}_1)^{1/2},\ldots,(\tilde{x}_{n_j})^{1/2}) \tag{6.16}$$

we consider the matrix

$$\tilde{X}F_j\tilde{X} \tag{6.17}$$

and realize that it satisfies the relation (use (6.13))

$$(\tilde{X}F_j\tilde{X})\tilde{X}^{-1}\tilde{x} = \tilde{X}\Phi(\bar{x})e^j = \Phi(\bar{x})\tilde{X}^{-1}\tilde{x}, \tag{6.18}$$

which means that the matrix (6.17) has the positive eigenvalue $\Phi(\bar{x})$. More information on the eigenvalues of this matrix can be drawn from the fact that in the case under consideration the matrix F_j has a d-dimensional kernel and because of its symmetry it has exactly $n_j - d$ real eigenvalues different from zero. By Sylvester's Inertia Theorem the same is true then for $\tilde{X}F_j\tilde{X}$, i.e. $\tilde{X}F_j\tilde{X}$ has eigenvalues, say,

$$\lambda_1 = \Phi(\bar{x}) > 0, \quad \lambda_i \neq 0, \quad i = 2,\dots,n_j - d. \tag{6.19}$$

Since $\tilde{X}F_j\tilde{X}$ is symmetric as well we can choose an orthogonal set $\{\tilde{y}_1,\dots,\tilde{y}_{n_j-d}\}$ of eigenvectors associated with the eigenvalues (6.19), i.e.

$$(\tilde{X}F_j\tilde{X})\tilde{y}_i = \lambda_i\tilde{y}_i \quad \text{for } i = 1,\dots,n_j - d, \tag{6.20}$$

where, in particular, $\tilde{y}_1 = \tilde{X}^{-1}\tilde{x}$. Later on we shall use the orthogonality relations

$$(\tilde{y}_1)^T\tilde{y}_i = \tilde{x}^T\tilde{X}^{-1}\tilde{y}_i = 0 \quad \text{for } i = 2,\dots,n_j - d,$$

or equivalently,

$$(e^j)^T\tilde{X}\tilde{y}_i = 0 \quad \text{for } i = 2,\dots,n_j - d. \tag{6.21}$$

Using that $\bar{x}$ is in Λ^{n-1}, i.e. that $(e^j)^T\tilde{x}$ equals 1, we get $n_j - d$ eigenvalue-eigenvector relations for $G_{11}(\bar{x})$. First, with (6.13)

$$G_{11}(\bar{x})\tilde{x} = \mathrm{diag}(\tilde{x})F_j\tilde{x} - 2\Phi(\bar{x})\tilde{x}(e^j)^T\tilde{x} =$$
$$\mathrm{diag}(\tilde{x})\Phi(\bar{x})e^j - 2\Phi(\bar{x})\tilde{x} = -\Phi(\bar{x})\tilde{x}$$

and, secondly, with $\tilde{X}^2 = \mathrm{diag}(\tilde{x})$, (6.20) and (6.21)

$$G_{11}(\bar{x})\tilde{X}\tilde{y}_i = \tilde{X}(\tilde{X}F_j\tilde{X})\tilde{y}_i - 2\Phi(\bar{x})\tilde{x}(e^j)^T\tilde{X}\tilde{y}_i =$$
$$\tilde{X}\lambda_i\tilde{y}_i = \lambda_i\tilde{X}\tilde{y}_i \quad \text{for } i = 2,\ldots,n_j - d.$$

Thus, $G_{11}(\bar{x})$ has the $n_j - d$ real nonzero eigenvalues $-\lambda_1 = -\Phi(\bar{x})$, λ_2, $\lambda_3, \ldots, \lambda_{n_j-d}$. This shows the validity of (6.15) and completes the proof of Lemma 6.4. ∎

Now the main result of this section follows rather easily.

<u>Theorem 6.2</u>: The ω-limit set $\Omega(x_o)$ of an arbitrary solution $x(t,x_o)$ of system (6.4) on Λ^{n-1} is either a continuum of exceptional points or a single (regular) point.

<u>Proof</u>: It is well known that $\Omega(x_o)$ as an ω-limit set is a continuum, i.e. a closed and connected set in Λ^{n-1}. From Lemma 6.2 we know that $\Omega(x_o)$ consists entirely of stationary solutions of (6.4). Hence, $\Omega(x_o)$ is either a continuum of exceptional points or it contains a regular point $\bar{x}$, say. In order to prove the theorem we have to show that in the latter case $x(t,x_o)$ converges to $\bar{x}$ as $t\to\infty$. Since $\bar{x}$ belongs to $\Omega(x_o)$, hypothesis (i) of Theorem 2.3 is fulfilled. Lemma 6.2 and Lemma 6.3 together say that there exists an $\mathbb{R}^n$-neighborhood U of $\bar{x}$ such that

$$\Omega(x_o) \cap U \subset E^{n-1} \cap U =: M$$

where M is a smooth d - dimensional manifold of stationary solutions of (6.4) ($d = d(\bar{x})$ is the defect of $\bar{x}$). Thus, hypothesis (ii) of Theorem 2.3 is satisfied. Lemma 6.4 finally shows that hypothesis (iii) of Theorem 2.3 is fulfilled as well. So we may apply this theorem and obtain herewith the desired relation $\lim_{t \to \infty} x(t,x_o) = \bar{x}$. ■

Theorem 6.2 extends the results in Aulbach and Hadeler [7] from the case $f_{jk} > 0$ to the general case $f_{jk} \geq 0$, $j,k = 1,\ldots,n$. We emphasize one point which seems to be of particular interest to biologists. The dynamical system described by equation (6.4) on the simplex Λ^{n-1} is called *cooperative* if the relative boundary of Λ^{n-1} is a repellor in the sense that any solution of (6.4) in the interior of Λ^{n-1} keeps a positive distance from the boundary of Λ^{n-1}. In biological terms this means that no allele becomes extinct if it is initially present. Since in the relative interior of Λ^{n-1} there are no exceptional points we immediately get from Theorem 6.2 the following result.

<u>Corollary</u>: If system (6.4) is cooperative, then any solution in the relative interior of Λ^{n-1} converges as time goes to infinity to a stationary solution all components of which are positive.

II. DIFFERENCE EQUATIONS

7. Basic Results

To begin with we state the basic theorem on the convergence problem in the discrete time case. Consider the difference equation

$$x(k+1) = f(x(k)), \tag{7.1}$$

where f is a C^3 - mapping of $\mathbb{R}^n$ into itself. The discrete time k ranges in $\mathbb{N}_0$, the set of nonnegative integers. We suppose that

(I) equation (7.1) has a stationary solution x^*, i.e. $f(x^*) = x^*$,

(II) there exists an m - dimensional C^3 - manifold M of stationary solutions of (7.1) with $x^* \in M$.

The main result of this book concerning the difference equation part is as follows.

Theorem 7.1: Under the above hypotheses (I) and (II) let x(k) be any solution of (7.1) for $k \geq 0$ with ω - limit set Ω. Suppose that

(i) there exists a point $x^* \in \Omega$.

(ii) there exists an $\mathbb{R}^n$-neighborhood U of x^* such that $\Omega \cap U \subset M$,

(iii) the matrix $f_x(x^*)$ has $n-m$ eigenvalues with moduli $\neq 1$,

(iv) $x(k)$ is bounded as $k \to \infty$,

(v) each ω-limit point of $x(k)$ is a stationary solution of (7.1).

Then $\lim_{k \to \infty} x(k) = x^*$.

Remark: Unlike the case of differential equations the solution paths of difference equations are not connected. This deficiency causes problems in our situation which will be overcome by means of the hypotheses (iv) and (v) of Theorem 7.1. Those hypotheses guarantee that

$$\lim_{k \to \infty} [x(k+1) - x(k)] = 0,$$

a relation which is sufficient for our purposes. In order to prove this limiting relation we suppose that the sequence $x(k+1) - x(k)$ does not converge to 0 as $k \to \infty$. This means that there exists a positive ε and a subsequence $\{k_\nu\}$ of $\{k\}$ such that

$$\| x(k_\nu+1) - x(k_\nu) \| \geq \varepsilon \text{ for all } \nu \in \mathbb{N}_0 .$$

Because of the boundedness assumption (iv) there exists a convergent subsequence $x(k_{\nu_\mu})$ of $x(k_\nu)$ whose limit x_∞, say, is by (v) a fixed point of f. This implies that $x(k_{\nu_\mu}+1) - x(k_{\nu_\mu}) = f(x(k_{\nu_\mu})) - x(k_{\nu_\mu})$ converges to $f(x_\infty) - x_\infty = 0$ as $\mu \to \infty$. This is a contradiction.

Proof of Theorem 7.1: Rather than considering solutions of the difference equation (7.1) we temporarily study the action of the map f on points of $\mathbb{R}^n$. Using a local coordinate chart (φ, U_φ) of M near x^* we may represent f in Euclidean (u,v,w) - coordinates as a C^3 - map $g = \varphi f \varphi^{-1}$ which is defined on an open neighborhood of the origin in $\mathbb{R}^n$. We arrange the local coordinates such that the set $\varphi(U_\varphi)$ of g - fixed points is, in the relative topology, an open neighborhood of the origin in the subspace of $\mathbb{R}^n$ with vanishing u - and v - coordinates. Thus, g may be assumed to have the form

$$g\begin{pmatrix} u \\ v \\ w \end{pmatrix} = \begin{pmatrix} A^- u + r_1(u,v,w) \\ A^+ v + r_2(u,v,w) \\ w + r_3(u,v,w) \end{pmatrix} \tag{7.2}$$

with

$$g\begin{pmatrix} 0 \\ 0 \\ w \end{pmatrix} = \begin{pmatrix} 0 \\ 0 \\ w \end{pmatrix}, \tag{7.3}$$

where $u \in \mathbb{R}^{n_-}$, $v \in \mathbb{R}^{n_+}$, $w \in \mathbb{R}^m$, $n_- + n_+ + m = n$. Here A^-, A^+ are constant matrices whose eigenvalues have moduli less than 1 and greater than 1, respectively. The functions $r_i(u,v,w)$, $i = 1,2,3$, are C^2 - nonlinearities of order $o(\|(u,v,w)\|)$ as $\|(u,v,w)\| \to 0$. The representation (7.2) and the identity (7.3) hold, of course, only on a suitably small neighborhood of the coordinate origin in $\mathbb{R}^n$. (Similar remarks concerning the local character of the analysis in this section will furthermore be omitted.) In terms of the nonlinearities the identity (7.3) reads

$$r_i(0,0,w) \equiv 0, \quad i = 1,2,3,$$

and this means that g may be written in the form

$$g\begin{pmatrix} u \\ v \\ w \end{pmatrix} = \begin{pmatrix} A^- u + A_1(u,v,w)u + B_1(u,v,w)v \\ A^+ v + A_2(u,v,w)u + B_2(u,v,w)v \\ w + A_3(u,v,w)u + B_3(u,v,w)v \end{pmatrix} \tag{7.4}$$

where

$$A_i(u,v,w) := \int_0^1 (r_i)_u(su,sv,w)\,ds\,, \quad i = 1,2,3,$$

$$B_i(u,v,w) := \int_0^1 (r_i)_v(su,sv,w)\,ds\,, \quad i = 1,2,3.$$

A_i, B_i are C^1 - matrices which vanish at $(u,v,w) = (0,0,0)$.

Our next aim is to get rid of the matrix $A_2(u,v,w)$ in the representation (7.4). The main tool we are going to use in this endeavor is a result on invariant manifolds due to Kirchgraber which can be viewed as the discrete analogue of Knobloch's result we have used in the continuous time case. For the sake of completeness we state a version of this result which is convenient for our purposes.

Theorem 7.2 (Kirchgraber and Stiefel [28, 12.Satz 1, Satz 3]): Consider a C^1 - map

$$P\begin{pmatrix} z \\ y \end{pmatrix} = \begin{pmatrix} F(z,y) \\ Ly + Y(z,y) \end{pmatrix},$$

where F and Y are defined for all $z \in \mathbb{R}^{n_-+m}$ and $y \in \mathbb{R}^{n_+}$, L is a constant $n_+ \times n_+$ matrix. Suppose that

(i) there exists a positive constant $\zeta < 1$ such that

$$\| L^{-1} \| \leq \zeta,$$

(ii) F and Y have derivatives which are uniformly continuous on $\mathbb{R}^{n_-+m} \times \mathbb{R}^{n_+}$ and there exist positive constants K_{zz}, K_{zy}, K_{yz}, K_{yy} such that

$$\| F_z(z,y) \| \leq K_{zz} \quad , \quad \| F_y(z,y) \| \leq K_{zy},$$
$$\| Y_z(z,y) \| \leq K_{yz} \quad , \quad \| Y_y(z,y) \| \leq K_{yy},$$

for all $(z,y) \in \mathbb{R}^{n_-+m} \times \mathbb{R}^{n_+}$,

(iii) there exists a positive constant $\rho \geq 1$ such that

$$\frac{1}{\zeta} - K_{yy} > \rho > K_{zz},$$

(iv) $$K_{yz}K_{zy} < \frac{1}{4}(\rho - K_{zz})(\frac{1}{\zeta} - K_{yy} - \rho).$$

Then there exists a C^1 - function $s: \mathbb{R}^{n_-+m} \to \mathbb{R}^{n_+}$ such that

(I) the manifold $N := \{(z,s(z)) : z \in \mathbb{R}^{n_-+m}\}$ is invariant under P, i.e. if $(z,y) \in N$ then the k - th iterate $P^k(z,y)$ of (z,y) under P belongs to N for all $k \in \mathbb{N}_0$,

(II) if for fixed (z,y) the y - component of $P^k(z,y)$ is bounded as $k \to \infty$ then (z,y) belongs to the manifold N.

Remarks: 1. For an application of this theorem it is crucial to note that the norm appearing in the theorem may be any vector norm and its corresponding operator norm.

2. The basic result corresponding to Theorem 7.2 can be found in Hartman [24, IX.Theorem 5.1]. Hartman's Theorem, however, does not meet our purposes because it requires a nonsingular linearization whereas our matrix A^- is allowed to have zero eigenvalues. This singular case will occur in the model equation of Section 11 (see p.97).

We want to apply Theorem 7.2 to our map (7.4). To this end we identify $y = v$, $z = (u,w)$, $L = A^+$. In order to verify the assumptions of

Theorem 7.2 we decompose the vector space $\mathbb{R}^n = \mathbb{R}^{n_-+n_++m}$ into the direct sum $\mathbb{R}^{n_-+m} \oplus \mathbb{R}^{n_+}$ and use the well known fact (see e.g. Stoer and Bulirsch [39, Theorem 6.9.2]) that for a given r×r matrix and any positive ε there exists a norm on $\mathbb{R}^r$ such that in this norm the matrix is smaller than its spectral radius plus ε. So we define a norm $\|\cdot\|_+$ on $\mathbb{R}^{n_+}$ such that

$$\|(A^+)^{-1}\|_+ \leq \frac{1}{2}[r((A^+)^{-1}) + 1]$$

where $r((A^+)^{-1})$ is the spectral radius of $(A^+)^{-1}$ which by assumption is less than 1. As ζ appearing in Theorem 7.2 we define

$$\zeta := \frac{1}{2}[r((A^+)^{-1}) + 1] ,$$

a number which is less than 1. By assumption, the spectral radius of $\mathrm{diag}(A^-, I_m)$ is 1 and thus there exists a norm $\|\cdot\|_-$ on $\mathbb{R}^{n_-+m}$ such that

$$\|\mathrm{diag}(A^-, I_m)\|_- \leq \frac{1}{2}(1 + \frac{1}{\zeta}) = \frac{(\zeta + 1)}{2\zeta} < \frac{1}{\zeta}.$$

By means of the norms $\|\cdot\|_+$ and $\|\cdot\|_-$ on $\mathbb{R}^{n_+}$ and $\mathbb{R}^{n_-+m}$, respectively, we define in an obvious way the norm

$$\|\cdot\| := \|\cdot\|_- \oplus \|\cdot\|_+$$

on $\mathbb{R}^n = \mathbb{R}^{n_-+m} \oplus \mathbb{R}^{n_+}$ implying the estimates

$$\|(A^+)^{-1}\| \leq \zeta, \quad \|\mathrm{diag}(A^-, I_m)\| \leq \frac{(\zeta + 1)}{2\zeta}.$$

Hence, in particular hypothesis (i) of Theorem 7.2 is satisfied. Next we alter the map (7.4) outside a ball B_σ of radius σ around $0 \in \mathbb{R}^n$ such that the new map satisfies the global boundedness assumptions (ii) of

Theorem 7.2. The nonlinearities in (7.4) vanish together with their first order derivatives at (0,0,0) and so, by a suitable choice of σ, we can make the global upper bounds K_{zy}, K_{yz}, K_{yy} for $F_y(z,y)$, $Y_z(z,y)$ and $Y_y(z,y)$, respectively, arbitrarily small. The bound K_{zz} for $F_z(z,y)$ may be made greater than but arbitrarily close to the bound $(\zeta+1)/2\zeta$ for $\mathrm{diag}(A^-, I_m)$. This fact is used in the following way. First of all, we choose σ so small that $K_{yy} < (1-\zeta)/4\zeta$, i.e.

$$\frac{1}{\zeta} - K_{yy} > \frac{3+\zeta}{4\zeta} > 1.$$

Keeping K_{yy} fixed we choose σ such that $K_{zz} < (\zeta+1)/2\zeta + (1-\zeta)/4\zeta = (\zeta+3)/4\zeta$, i.e.

$$K_{zz} < \frac{3+\zeta}{4\zeta} < \frac{1}{\zeta} - K_{yy}$$

and this means that hypothesis (iii) of Theorem 7.2 is fulfilled. Next we choose any ρ in the open interval $(K_{zz}, \frac{1}{\zeta} - K_{yy})$, keep K_{zz} and K_{yy} fixed and take σ so small that

$$K_{yz}K_{zy} < \frac{1}{4}(\rho - K_{zz})(\frac{1}{\zeta} - K_{yy} - \rho).$$

Thus, all hypotheses of Theorem 7.2 are verified for a map which coincides with (7.4) on a small ball B_σ around $0 \in \mathbb{R}^n$. Theorem 7.2 says then that our map (7.4) has a locally invariant manifold

$$v = s(u,w). \tag{7.5}$$

Furthermore, if a sequence $g^k(u_o,v_o,w_o)$ of g-iterates remains in B_σ for all $k \in \mathbb{N}_o$, then

$$v_o = s(u_o,w_o).$$

The particular form (7.4) shows that $g^k(0,0,w) = (0,0,w)$ for all $k \in \mathbb{N}_0$ and therefore

$$s(0,w) \equiv 0 \text{ for } \|w\| \text{ suitably small.}$$

The invariance of the manifold $v = s(u,w)$ under the map g is reflected by the identity

$$\begin{aligned} &A^+ s(u,w) + A_2(u,s(u,w),w)u + B_2(u,s(u,w),w)s(u,w) \equiv \\ &s(A^- u + A_1(u,s(u,w),w)u + B_1(u,s(u,w),w)s(u,w), \\ &w + A_3(u,s(u,w),w)u + B_3(u,s(u,w),w)s(u,w)) \end{aligned} \tag{7.6}$$

which is valid for $\|u\|$ and $\|w\|$ small. These facts are now used to introduce new curvilinear coordinates $(\bar{u},\bar{v},\bar{w})$ near $0 \in \mathbb{R}^n$ via

$$\begin{aligned} \bar{u} &= u \\ \bar{v} &= v - s(u,w) \\ \bar{w} &= w . \end{aligned}$$

In this coordinate system the map under consideration assumes the form

$$h\begin{pmatrix}\bar{u}\\ \bar{v}\\ \bar{w}\end{pmatrix} := g\begin{pmatrix}\bar{u}\\ \bar{v}+s(\bar{u},\bar{w})\\ \bar{w}\end{pmatrix} - \begin{pmatrix}0\\ s(g_1(\bar{u},\bar{v}+s(\bar{u},\bar{w}),\bar{w}),g_3(\bar{u},\bar{v}+s(\bar{u},\bar{w}),\bar{w}))\\ 0\end{pmatrix},$$

where $g = (g_1,g_2,g_3)^T$. Supposing the corresponding partition $(h_1,h_2,h_3)^T$ for the function h the explicit form of h_1 is easily derived to be

$$h_1\begin{pmatrix}\bar{u}\\ \bar{v}\\ \bar{w}\end{pmatrix} = A^-\bar{u} + A_1(\bar{u},\bar{v}+s(\bar{u},\bar{w}),\bar{w})\bar{u} + B_1(\bar{u},\bar{v}+s(\bar{u},\bar{w}),\bar{w})(\bar{v}+s(\bar{u},\bar{w})).$$

Recalling the identity $s(0,w) \equiv 0$ the function h_1 may be written as

$$h_1\begin{pmatrix}\bar{u}\\ \bar{v}\\ \bar{w}\end{pmatrix} = A^-\bar{u} + \bar{A}_1(\bar{u},\bar{v},\bar{w})\bar{u} + \bar{B}_1(\bar{u},\bar{v},\bar{w})\underset{\sim}{v}$$

with continuous matrices $\bar{A}_1$, $\bar{B}_1$ which vanish at $(\bar{u},\bar{v},\bar{w}) = (0,0,0)$. Analogous considerations yield for the third component of h the representation

$$h_3\begin{pmatrix}\bar{u}\\ \bar{v}\\ \bar{w}\end{pmatrix} = \bar{w} + \bar{A}_3(\bar{u},\bar{v},\bar{w})\bar{u} + \bar{B}_3(\bar{u},\bar{v},\bar{w})\bar{v}.$$

For the crucial second component we get

$$h_2\begin{pmatrix}\bar{u}\\ \bar{v}\\ \bar{w}\end{pmatrix} = \overset{+}{A}\bar{v} + B_2(\bar{u},\bar{v}+s(\bar{u},\bar{w}),\bar{w})\bar{v} + R(\bar{u},\bar{v},\bar{w})$$

where $R(\bar{u},\bar{v},\bar{w})$ is an abbreviation for the expression

$$\begin{aligned}&A^{+}s(\bar{u},\bar{w}) + A_2(\bar{u},\bar{v}+s(\bar{u},\bar{w}),\bar{w})\bar{u} + B_2(\bar{u},\bar{v}+s(\bar{u},\bar{w}),\bar{w})s(\bar{u},\bar{w}) - \\ &s(\bar{A}\bar{u}+A_1(\bar{u},\bar{v}+s(\bar{u},\bar{w}),\bar{w})\bar{u}+B_1(\bar{u},\bar{v}+s(\bar{u},\bar{w}),\bar{w})(\bar{v}+s(\bar{u},\bar{w})), \\ &\bar{w}+A_3(\bar{u},\bar{v}+s(\bar{u},\bar{w}),\bar{w})\bar{u}+B_3(\bar{u},\bar{v}+s(\bar{u},\bar{w}),\bar{w})(\bar{v}+s(\bar{u},\bar{w}))).\end{aligned}$$

Because of (7.6), the identity $R(\bar{u},0,\bar{w}) \equiv 0$ holds true and thus $R(\bar{u},\bar{v},\bar{w})$ has the form $\tilde{A}_2(\bar{u},\bar{v},\bar{w})\bar{v}$ with a continuous matrix $\tilde{A}_2$ which vanishes at $(\bar{u},\bar{v},\bar{w}) = (0,0,0)$.

Altogether, the map under consideration, called h in $(\bar{u},\bar{v},\bar{w})$ - coordinates, has the form

$$h\begin{pmatrix}\bar{u}\\ \bar{v}\\ \bar{w}\end{pmatrix} = \begin{pmatrix}\bar{A}\bar{u} + \bar{A}_1(\bar{u},\bar{v},\bar{w})\bar{u} + \bar{B}_1(\bar{u},\bar{v},\bar{w})\bar{v}\\ A^{+}\bar{v} \qquad\qquad + \bar{B}_2(\bar{u},\bar{v},\bar{w})\bar{v}\\ \bar{w} + \bar{A}_3(\bar{u},\bar{v},\bar{w})\bar{u} + \bar{B}_3(\bar{u},\bar{v},\bar{w})\bar{v}\end{pmatrix} \tag{7.7}$$

with continuous matrices $\bar{A}_i$, $i = 1,3$, $\bar{B}_j$, $j = 1,2,3$, which vanish at $(\bar{u},\bar{v},\bar{w}) = (0,0,0)$. This map now exhibits the kind of decoupling and linearity which allows us to complete the proof of Theorem 7.1.

The situation of the theorem, stated in the $(\bar{u},\bar{v},\bar{w})$ - coordinate system is as follows. There exists a sequence (not necessarily a sequence of iterates yet) $(\bar{u}(k),\bar{v}(k),\bar{w}(k))$, a σ - neighborhood B_σ of the coordinate origin in $\mathbb{R}^n$ and a sequence of integers k_ν, $\nu \in \mathbb{N}_0$, with $k_\nu \to \infty$ as $\nu \to \infty$ such that

$$\lim_{\nu \to \infty} (\bar{u}(k_\nu),\bar{v}(k_\nu),\bar{w}(k_\nu)) = (0,0,0). \tag{7.8}$$

Furthermore,

$$\begin{array}{l} \text{each } \omega \text{ - limit point } (\bar{u}_\infty,\bar{v}_\infty,\bar{w}_\infty) \text{ of} \\ (\bar{u}(k),\bar{v}(k),\bar{w}(k)),\ k \in \mathbb{N}_0,\ \text{in } B_\sigma \\ \text{satisfies } \bar{u}_\infty = 0,\ \bar{v}_\infty = 0, \end{array} \tag{7.9}$$

and

$$\lim_{k \to \infty} (\bar{u}(k+1) - \bar{u}(k),\bar{v}(k+1) - \bar{v}(k),\bar{w}(k+1) - \bar{w}(k)) = (0,0,0). \tag{7.10}$$

(For (7.10) see the remark to Theorem 7.1.) Moreover, for each ν there exists a set S_ν of consecutive integers with $k_\nu \in S_\nu$ such that $(\bar{u}(k),\bar{v}(k),\bar{w}(k))$ is a sequence of h - iterates in B_σ as long as k is in S_ν, $\nu \in \mathbb{N}_0$. In order to prove the convergence of $(\bar{u}(k),\bar{v}(k),\bar{w}(k))$ we suppose to the contrary that

$$\begin{array}{l} (\bar{u}(k),\bar{v}(k),\bar{w}(k)) \text{ does not con-} \\ \text{verge to } (0,0,0) \text{ as } k \to \infty. \end{array} \tag{7.11}$$

More explicitly, this assumption can be stated in the following way. There exists a positive constant ρ ($\leq \sigma$) and a sequence of sets

$$T_\nu := \{k_\nu, k_\nu+1, \ldots, K_\nu\} \subset S_\nu,$$
$$k_\nu+1 \leq K_\nu, \quad \nu \in \mathbb{N}_0,$$

of consecutive integers such that

$$(\bar{u}(k),\bar{v}(k),\bar{w}(k)) \in B_\rho \text{ for all } k \in \bigcup_{\nu\in\mathbb{N}_0} T_\nu,$$

and

$$(\bar{u}(K_\nu),\bar{v}(K_\nu),\bar{w}(K_\nu)) \in B_\rho \setminus B_{\rho/2} \text{ for all } \nu \in \mathbb{N}_0. \tag{7.12}$$

Without loss of generality the constant ρ may be chosen so small that

$$\| \bar{A}_i(\bar{u}(k),\bar{v}(k),\bar{w}(k)) \| \leq \frac{1-\lambda}{2\gamma} \text{ for all } k \in \bigcup_{\nu\in\mathbb{N}_0} T_\nu, \; i = 1,3,$$

$$\| \bar{B}_j(\bar{u}(k),\bar{v}(k),\bar{w}(k)) \| \leq \frac{1-\lambda}{2\gamma} \text{ for all } k \in \bigcup_{\nu\in\mathbb{N}_0} T_\nu, \; j = 1,2,3,$$

where $\gamma \geq 1$ and $\lambda \in (0,1)$ are constants such that (don't confuse the letter "l" with the numeral "1")

$$\| (A^-)^{k-l} \| \leq \gamma\lambda^{k-l} \text{ for all } k \geq l \geq 0,$$

$$\| (A^+)^{k-l} \| \leq \gamma\lambda^{l-k} \text{ for all } l \geq k \geq 0.$$

This choice of γ and λ is possible because of the spectral properties of A^- and A^+. Thus, for each $\nu \in \mathbb{N}_0$, we may consider $(\bar{u}(k),\bar{v}(k),\bar{w}(k))$ as a solution of the difference equation

$$\begin{aligned}
\bar{u}(k+1) &= A^-\bar{u}(k) + \bar{A}_1(\bar{u}(k),\bar{v}(k),\bar{w}(k))\bar{u}(k) + \bar{B}_1(\bar{u}(k),\bar{v}(k),\bar{w}(k))\bar{v}(k)\\
\bar{v}(k+1) &= A^+\bar{v}(k) + \bar{B}_2(\bar{u}(k),\bar{v}(k),\bar{w}(k))\bar{v}(k)\\
\bar{w}(k+1) &= \bar{w}(k) + \bar{A}_3(\bar{u}(k),\bar{v}(k),\bar{w}(k))\bar{u}(k) + \bar{B}_3(\bar{u}(k),\bar{v}(k),\bar{w}(k))\bar{v}(k)
\end{aligned}$$

on T_ν. Now we may apply Lemma B.6, Appendix B, providing the estimate

$$\|\bar{w}(K_\nu)\| \leq \|\bar{w}(k_\nu)\| + \|\bar{u}(k_\nu)\| + 2\|\bar{v}(K_\nu)\| \text{ for all } \nu \in \mathbb{N}_0. \tag{7.13}$$

Because of (7.12), the sequence $(\bar{u}(K_\nu),\bar{v}(K_\nu),\bar{w}(K_\nu))$, $\nu \in \mathbb{N}_0$, is bounded and therefore has a convergent subsequence $(\bar{u}(K_{\nu_\mu}),\bar{v}(K_{\nu_\mu}),\bar{w}(K_{\nu_\mu}))$ with limit $(\bar{u}_\infty, \bar{v}_\infty, \bar{w}_\infty)$, say. Employing (7.9) we get

$$\lim_{\mu\to\infty} \bar{u}(K_{\nu_\mu}) = \bar{u}_\infty = 0,$$
$$\lim_{\mu\to\infty} \bar{v}(K_{\nu_\mu}) = \bar{v}_\infty = 0, \tag{7.14}$$

and with (7.12)

$$\| \lim_{\mu\to\infty} \bar{w}(K_{\nu_\mu}) \| = \|\bar{w}_\infty\| \geq \frac{\rho}{2} > 0. \tag{7.15}$$

On the other hand, combining (7.8), (7.13) and (7.14) implies the validity of the relation

$$\lim_{\mu\to\infty} \| \bar{w}(K_{\nu_\mu}) \| = \|\bar{w}_\infty\| = 0$$

which contradicts (7.15). Hence, statement (7.11) leads to a contradiction and Theorem 7.1 is proved. ∎

8. Asymptotic Phase

In this section we treat the discrete form of the asymptotic phase problem that has been considered in the continuous time setting in Section 3. Here, the underlying equation is of the form

$$x(k+1) = f(x(k)) \tag{8.1}$$

where $f: \mathbb{R}^n \to \mathbb{R}^n$ is supposed to be of class C^3. The basic assumption of this section is that system (8.1) admits a normally hyperbolic compact manifold M of stationary solutions, i.e.

(i) f has a compact C^3 - manifold M of fixed points, $m := \dim M$,

(ii) for each fixed point $\bar{x} \in M$ the Jacobian $f_x(\bar{x})$ has $n-m$ eigenvalues with moduli different from 1.

We note in passing that $f_x(\bar{x})$ is allowed to have zero eigenvalues. This means that the well known theory of invariant manifolds for diffeomorphisms due to Hirsch, Pugh and Shub [27] and others cannot be employed. The particular case of diffeomorphisms will be treated separately in Section 10.

As in the continuous time case we may draw an immediate conclusion from our basic convergence result in Theorem 7.1.

Proposition 8.1: If some solution x(k) of (8.1) with the property that $\lim_{k\to\infty}[x(k+1)-x(k)] = 0$ satisfies the relation

$$\lim_{k\to\infty} \operatorname{dist}(x(k),M) = 0,$$

then there exists a point $x^* \in M$ such that

$$\lim_{k\to\infty} x(k) = x^*.$$

In this section we show that the assumption that x(k) tends to M can be weakened.

Theorem 8.1: Under the above hypotheses there exists an $\mathbb{R}^n$- neighborhood N of M with the following property: if for some $k_o \in \mathbb{N}_o$ a solution x(k) of (8.1) with $\lim_{k\to\infty}[x(k+1)-x(k)] = 0$ remains in N for all $k \geq k_o$, then there exists a point $x^* \in M$ such that

$$\lim_{k \to \infty} x(k) = x^*.$$

Proof: We suppose to the contrary of Theorem 8.1 that in any neighborhood of M there exists a positive semiorbit of (8.1) which does not converge to a point on M. This particularly means that there exists a sequence $x_\nu(k)$ of solutions of (8.1) with

$$\lim_{k \to \infty} [x_\nu(k+1) - x_\nu(k)] = 0, \quad \nu \in \mathbb{N}_0, \tag{8.2}$$

and a sequence of integers k_ν such that

$$\begin{aligned} &x_\nu(k) \in N_\nu \quad \text{for all } k \geq k_\nu, \quad \nu \in \mathbb{N}_0 \quad \text{where} \\ &N_\nu := \{x \in \mathbb{R}^n : \operatorname{dist}(x,M) \leq \tfrac{1}{\nu}\}. \end{aligned} \tag{8.3}$$

Since M is compact so is any N_ν and thus any solution $x_\nu(k)$ has an ω-limit point x_ν^∞, say, in N_ν. The sequence x_ν^∞ of ω-limit points is bounded and so we may choose a convergent subsequence $x_{\nu_\mu}^\infty$, denoting its limit by

$$x_\infty^\infty := \lim_{\mu \to \infty} x_{\nu_\mu}^\infty \in M.$$

As in the previous section we now introduce local coordinates such that system (8.1) near x_∞^∞ may be written as

$$\begin{aligned} u(k+1) &= A^- u(k) + A_1(u(k),v(k),w(k))u(k) + B_1(u(k),v(k),w(k))v(k) \\ v(k+1) &= A^+ v(k) \qquad\qquad\qquad\qquad\qquad\qquad\quad + B_2(u(k),v(k),w(k))v(k) \\ w(k+1) &= \quad w(k) + A_3(u(k),v(k),w(k))u(k) + B_3(u(k),v(k),w(k))v(k) \end{aligned} \tag{8.4}$$

where x_∞^∞ corresponds to the (u,v,w)-coordinate origin. The matrices A^-, A^+ have eigenvalues with moduli less than one and greater than one, respectively, i.e. there exist constants $\gamma \geq 1$ and $0 < \lambda < 1$ such that

$$\|(A^-)^{k-l}\| \leq \gamma\,\lambda^{k-l} \quad \text{for all } k \geq l \geq 0,$$
$$\|(A^+)^{k-l}\| \leq \gamma\,\lambda^{l-k} \quad \text{for all } l \geq k \geq 0. \tag{8.5}$$

The matrices A_i, $i = 1,3$, B_j, $j = 1,2,3$, are continuous functions of (u,v,w) which vanish at $(0,0,0)$. Thus, we may choose a positive constant σ such that

$$\|A_i(u,v,w)\| \leq \frac{1-\lambda}{2\gamma}, \quad \|B_j(u,v,w)\| \leq \frac{1-\lambda}{2\gamma}$$
$$\text{if } \|(u,v,w)\| \leq \sigma. \tag{8.6}$$

The role of σ is described in the following lemma.

<u>Lemma 8.1</u>: If for some $k_o \in \mathbb{N}_o$ a solution $(u_o(k),v_o(k),w_o(k))$ of (8.4) satisfies $\|(u_o(k),v_o(k),w_o(k))\| \leq \sigma$ for all $k \geq k_o$, then there exists a point $(0,0,w^*) \in B_\sigma$ such that

$$\lim_{k\to\infty} (u_o(k),v_o(k),w_o(k)) = (0,0,w^*).$$

<u>Proof of Lemma 8.1</u>: The sequence $(u_o(k),v_o(k),w_o(k))$, $k \geq k_o$, may be considered as a solution of the linear system

$$\begin{aligned} u(k+1) &= [A^- + C_1^o(k)]\,u(k) + D_1^o(k)\,v(k) \\ v(k+1) &= [A^+ + D_2^o(k)]\,v(k) \\ w(k+1) &= w(k) + C_3^o(k)\,u(k) + D_3^o(k)\,v(k), \end{aligned} \tag{8.7}$$

where

$$C_i^o(k) := A_i(u_o(k),v_o(k),w_o(k)), \quad i = 1,3,$$
$$D_j^o(k) := B_j(u_o(k),v_o(k),w_o(k)), \quad j = 1,2,3.$$

The assumptions of the lemma together with (8.6) imply that

$$\| C_i^o(k) \| \leq \frac{1-\lambda}{2\gamma}, \quad \| D_j^o(k) \| \leq \frac{1-\lambda}{2\gamma} \text{ for all } k \geq k_o,$$

and we may apply the Corollary to Lemma B.5, Appendix B. Since system (8.7) is decoupled we may investigate the three components of the solution $(u_o(k), v_o(k), w_o(k))$ separately.

First we show that

$$v_o(k) = 0 \text{ for all } k \geq k_o. \tag{8.8}$$

Since the principal fundamental matrix $\Phi(k,l)$ of part (ii) of Lemma B.5 in our case equals $(A^+)^{k-l}$ and since the inequalities $\gamma \geq 1$ and $0 < \lambda < 1$ imply that

$$\frac{1-\lambda}{2\gamma} < \frac{1-\lambda}{\lambda\gamma(1+\lambda)}$$

we get for the principal fundamental matrix $\hat{\Phi}(k,l)$ of the system $v(k+1) = [A^+ + D_2^o(k)]\, v(k)$ the estimate

$$\| \hat{\Phi}(k,l) \| \leq \gamma\left(\frac{1+\lambda}{2}\right)^{l-k} \text{ for all } l \geq k+1 \geq k_o+1.$$

The identity $\hat{\Phi}(k_o,k)v_o(k) = v_o(k_o)$ implies that

$$\| v_o(k_o) \| \leq \| v_o(k) \| \, \gamma\left(\frac{1+\lambda}{2}\right)^{k-k_o} \text{ for all } k \geq k_o+1$$

and this means that $\| v_o(k) \|$ is unbounded as $k \to \infty$ unless $v_o(k_o)$ is zero. The boundedness assumption of Lemma 8.1 thus gives the validity of (8.8).

Next we consider the sequence $u_o(k)$, $k \geq k_o$. Because of (8.7), (8.8) this sequence satisfies the relation

$$u_o(k+1) = [A^- + C_1^o(k)]u_o(k) \text{ for all } k \geq k_o$$

and we may apply part (i) of the Corollary to Lemma B.5 which immediately implies that

$$\| u_o(k) \| \leq \| u_o(k_o) \| \gamma (\frac{1+\lambda}{2})^{k-k_o} \text{ for all } k \geq k_o+1. \tag{8.9}$$

Finally, $w_o(k)$ is, for $k \geq k_o$, a solution of the equation

$$w(k+1) = w(k) + C_3^o(k)u_o(k). \tag{8.10}$$

Similarly to the continuous time case we show that the sequence $w_o(k)$ converges as $k \to \infty$. Because of (8.9) (note that $(1+\lambda)/2 < 1$) we get

$$\sum_{\kappa=k_o}^{\infty} C_3^o(\kappa)u_o(\kappa) < \infty,$$

and thus the sequence

$$w_1(k) := - \sum_{\kappa=k}^{\infty} C_3^o(\kappa)u_o(\kappa), \; k \geq k_o,$$

is a solution of (8.10) which moreover converges to zero as $k \to \infty$. On the other hand, any solution of (8.10) can be represented as

$$w_1(k) + \tilde{w}, \; k \geq k_o,$$

where $\tilde{w}$ is constant (in fact, a solution of the homogeneous equation $w(k+1) = w(k)$ associated with (8.10)) and so there exists a constant w^* such that

$$w_o(k) = w_1(k) + w^*,\ k \geq k_o.$$

This shows that

$$\lim_{k \to \infty} w_o(k) = w^*$$

and the proof of the lemma is complete. ∎

With Lemma 8.1 at hand we are able to complete the proof of Theorem 8.1. The situation in (u,v,w) - coordinates is the following. In any neighborhood of the coordinate origin there exists an ω - limit point of a solution of (8.4). Because of (8.3), this solution is supposed to creep along the space $u = 0$, $v = 0$ of stationary solutions as long as it is in B_σ. To be more explicit, we may choose a positive ε,

$$\varepsilon < \frac{\sigma}{12}, \tag{8.11}$$

with the following property. There exists a sequence $(u_\varepsilon(k), v_\varepsilon(k), w_\varepsilon(k))$, $k \in \mathbb{N}_o$, with

$$\lim_{k \to \infty} [(u_\varepsilon(k+1), v_\varepsilon(k+1), w_\varepsilon(k+1)) - (u_\varepsilon(k), v_\varepsilon(k), w_\varepsilon(k))] = (0,0,0) \tag{8.12}$$

which has an ω - limit point in B_ε, i.e. there exists a sequence k_ν of integers, $k_\nu \to \infty$ as $\nu \to \infty$, such that

$$\| \lim_{\nu \to \infty} (u_\varepsilon(k_\nu), v_\varepsilon(k_\nu), w_\varepsilon(k_\nu)) \| < \varepsilon. \tag{8.13}$$

Furthermore, there exists a sequence of sets J_ν, where each J_ν consists of consecutive integers with $k_\nu \in J_\nu$, such that

$$(u_\varepsilon(k), v_\varepsilon(k), w_\varepsilon(k)) \text{ is a solution of (8.4) on } J_\nu,\ \nu \in \mathbb{N}_0. \tag{8.14}$$

Moreover, this solution has the property that

$$\|u_\varepsilon(k)\| \leq \varepsilon,\ \|v_\varepsilon(k)\| \leq \varepsilon,\ \|w_\varepsilon(k)\| \leq \varepsilon \text{ for all } k \in J_\nu,\ \nu \in \mathbb{N}_0, \tag{8.15}$$

and finally it is supposed in this proof that

$$(u_\varepsilon(k), v_\varepsilon(k), w_\varepsilon(k)) \text{ does not converge to any point of the form } (0,0,w). \tag{8.16}$$

As in the differential equation part we have two cases.

Case 1: $(u_\varepsilon(k), v_\varepsilon(k), w_\varepsilon(k))$ belongs to B_σ for all integers $k \geq \bar{k}$, for some $\bar{k}$. Then, by Lemma 8.1 the sequence $(u_\varepsilon(k), v_\varepsilon(k), w_\varepsilon(k))$, $k \geq \bar{k}$, converges to a point $(0,0,w^*)$ as $k \to \infty$ and this is a contradiction to the above statement (8.16).

Case 2: $(u_\varepsilon(k), v_\varepsilon(k), w_\varepsilon(k))$ leaves B_σ again and again. Because of the relation (8.12) this means that there exists a sequence $K_\nu \in J_\nu$, $\nu \in \mathbb{N}_0$, such that

$$\frac{\sigma}{2} \leq \|(u_\varepsilon(K_\nu), v_\varepsilon(K_\nu), w_\varepsilon(K_\nu))\| \leq \sigma \text{ for all } \nu \in \mathbb{N}_0.$$

With (8.15) we infer that (using a suitable norm)

$$\|w_\varepsilon(K_\nu)\| \geq \frac{\sigma}{2} - 2\varepsilon > 0 \text{ for all } \nu \in \mathbb{N}_0. \tag{8.17}$$

On the other hand, $(u_\varepsilon(k), v_\varepsilon(k), w_\varepsilon(k))$ may be considered as a solution of the linear system

$$u(k+1) = [A^- + C_1^\varepsilon(k)]u(k) + D_1^\varepsilon(k)v(k)$$
$$v(k+1) = [A^+ + D_2^\varepsilon(k)]v(k)$$
$$w(k+1) = w(k) + C_3^\varepsilon(k)u(k) + D_3^\varepsilon(k)v(k)$$

on each set $\{k_\nu,\dots,K_\nu\}$ of consecutive integers, where the matrices C_i^ε, D_j^ε are defined to be

$$C_i^\varepsilon(k) := A_i(u_\varepsilon(k),v_\varepsilon(k),w_\varepsilon(k)),\quad i=1,3,$$
$$D_j^\varepsilon(k) := B_j(u_\varepsilon(k),v_\varepsilon(k),w_\varepsilon(k)),\quad j=1,2,3.$$

The above choice of σ guarantees that

$$\| C_i^\varepsilon(k)\| \leq \frac{1-\lambda}{2\gamma},\quad \| D_j^\varepsilon(k)\| \leq \frac{1-\lambda}{2\gamma} \text{ on each set } \{k_\nu,\dots,K_\nu\},\ \nu \in \mathbb{N}_0,$$

and this means that we may apply Lemma B.6, Appendix B. This yields the estimates

$$\| w_\varepsilon(K_\nu)\| \leq \| w_\varepsilon(k_\nu)\| + \| u_\varepsilon(k_\nu)\| + 2\| v_\varepsilon(K_\nu)\| \quad \text{for all } \nu \in \mathbb{N}_0.$$

Using the inequalities (8.13) and (8.15) we may conclude that

$$\| w_\varepsilon(K_\nu)\| \leq 4\varepsilon \text{ for } \nu \text{ sufficiently large,}$$

and this contradicts (8.17) by the choice (8.11) of ε.

Thus, in each of the two cases considered we get a contradiction and so the theorem is proved. ∎

9. Saddle Point Property

This section presents the discrete analogue of the results of Section 4. Again, the result on the saddle point property follows easily from the main result of the asymptotic phase section.

A fixed point x_o of a map f or, equivalently, a stationary solution x_o of a difference equation

$$x(k+1) = f(x(k)) \tag{9.1}$$

is said to have the *saddle point property* if there exists a neighborhood N of x_o such that any positive semiorbit in N lies on the stable manifold of x_o. It is well known that x_o has this property if the Jacobian $f_x(x_o)$ is nonsingular and has no eigenvalues on the unit circle in the complex plane. Of course, a nonisolated fixed point of f fails to have the saddle point property and so does any point on a manifold M of fixed points. Turning to the manifold M as a whole, however, there is a natural generalization of the saddle point property. To prove the corresponding result is the aim of this section.

We impose the same conditions on system (9.1) as in the previous section, namely

(i) f has a compact C^3 - manifold M of fixed points, m := dim M,

(ii) for each point $\bar{x} \in M$ the matrix $f_x(\bar{x})$ has n - m eigenvalues with moduli different from 1.

Theorem 9.1: Under the above hypotheses there exists an $\mathbb{R}^n$ - neighborhood N of M with the following property: if for some $k_o \in \mathbb{N}_o$ a solution $x(k)$ of (9.1) with $\lim_{k\to\infty}[x(k+1)-x(k)] = 0$ remains in N for all $k \ge k_o$, then $x(k)$ lies on the stable manifold of some point on M.

Proof: Theorem 8.1 guarantees the existence of a point $x^* \in M$ such that

$$\lim_{k\to\infty} x(k) = x^*.$$

As in the continuous time case we can conclude from this relation that $x(k)$ lies on the stable manifold of f through x^*. The argument is this: the solution $x(k)$ eventually stays arbitrarily close to x^* and consequently $x(k)$ lies on a center-stable manifold of x^*. This manifold, however, is fibred by stable manifolds associated with the stationary solutions on M. Thus, $x(k)$ lies on the stable manifold through x^* and the proof of Theorem 9.1 is complete. ■

10. Equations with a Nonstationary Invariant Manifold

As in the continuous time case the analysis of the convergence problem being treated in this book becomes simpler if the given equation has solutions which are defined on $(-\infty,\infty)$. This is the case e.g. if for the given equation (7.1) the linearization of the right hand side around the stationary solution x^* is nonsingular. On the other hand, in this case one can drop the assumption that there exists a manifold of stationary solutions. To wit, the stationary manifold can be replaced by a center manifold whose flow is stable relative to x^*. This will be shown in the present section.

We investigate the difference equation

$$x(k+1) = f(x(k)), \; f(0) = 0, \tag{10.1}$$

where $f: \mathbb{R}^n \to \mathbb{R}^n$ is three times continuously differentiable. The crucial hypothesis of this section which guarantees the unique backward continuation of the solutions of (10.1) near $x = 0$ is that the Jacobian

$$f_x(0) \text{ is nonsingular.} \tag{10.2}$$

Hence, f is a diffeomorphism of a neighborhood U of $0 \in \mathbb{R}^n$ onto the neighborhood f(U) of $0 \in \mathbb{R}^n$. Otherwise $f_x(0)$ is arbitrary, i.e. $f_x(0)$ may have eigenvalues inside, outside and on the unit circle in the complex plane. The respective numbers of those eigenvalues are denoted by n_-, n_+, n_o. Without loss of generality we may write (10.1) in the form

$$\begin{aligned} u(k+1) &= A^- u(k) + r_1(u(k),v(k),w(k)) \\ v(k+1) &= A^+ v(k) + r_2(u(k),v(k),w(k)) \\ w(k+1) &= A^o w(k) + r_3(u(k),v(k),w(k)), \end{aligned} \tag{10.3}$$

where A^-, A^+, A^o are matrices whose eigenvalues have moduli less than, greater than or equal to 1, respectively. System (10.3) has local center manifolds near the coordinate origin. The notion of an invariant manifold is not so common for difference equations. However, via the advance map associated with a difference equation, one can borrow this concept from the well known invariant manifold theory for maps (see e.g. Hartman [24, Chapter IX] and Hirsch, Pugh and Shub [27]). We suppose that the zero solution of (10.1) is stable for both time directions with respect to the flow on a center manifold. Since any solution of (10.1) which remains for all integers k sufficiently close to $x = 0$ lies on any center manifold through $x = 0$, there exists only one local center manifold through $x = 0$ under this stability hypothesis.

As it was true for the two basic results (Theorem 2.1 and Theorem 7.1) on our convergence problem, also here the theorem on the discrete time case requires two hypotheses more than the corresponding result (Theorem 5.1) for the continuous time case.

<u>Theorem 10.1</u>: Let $x(k)$ be any solution of (10.1), (10.2) with ω-limit set Ω. Suppose that

(i) $0 \in \Omega$,

(ii) the zero solution of (10.1) is stable in either time direction with respect to the (uniquely determined) local center manifold C through $x = 0$,

(iii) there exists a neighborhood U of $0 \in \mathbb{R}^n$ such that $\Omega \cap U \subset C$,

(iv) $x(k)$ is bounded as $k \to \infty$,

(v) $\lim_{k \to \infty} [x(k+1) - x(k)] = 0$.

Then $\lim_{k \to \infty} x(k) = 0$.

As mentioned in the introduction this theorem can be proved by an application of Hirsch, Pugh and Shub [27, Theorem 4.1]. The verification of the hypotheses of this theorem, however, would require the use of some concepts from differential topology. We want to give another proof of Theorem 10.1 which is more in the spirit of this book. This proof is based on a result on topological equivalence due to Kirchgraber. This result which we state now can be viewed as the discrete analogue of Palmer's Theorem in [36], [37] we referred to at the end of Section 2.

Theorem 10.2 (Kirchgraber [29, 5.Theorem 1]): For the homeomorphism

$$P: \begin{pmatrix} u \\ v \\ w \end{pmatrix} \rightarrow \begin{pmatrix} Au + R_1(u,v,w) \\ Bv + R_2(u,v,w) \\ \omega(w) + R_3(u,v,w) \end{pmatrix} \tag{10.4}$$

and its inverse

$$P^{-1}: \begin{pmatrix} u \\ v \\ w \end{pmatrix} \rightarrow \begin{pmatrix} A^{-1}u + S_1(u,v,w) \\ B^{-1}v + S_2(u,v,w) \\ \omega^{-1}(w) + S_3(u,v,w) \end{pmatrix}$$

suppose that

(i) ω and ω^{-1} are Lipschitzian with Lipschitz constant λ,

(ii) R_1, R_2, R_3, S_1, S_2, S_3 are Lipschitzian with Lipschitz constant κ,

(iii) R_2 and S_1 are bounded,

(iv) $\|Au\| \leq \theta\|u\|$, $\|B^{-1}v\| \leq \theta\|v\|$ for some $\theta > 0$.

Then, if $\theta < 1$ and $\lambda\theta < 1$, there exists a $\kappa_0 > 0$ such that for $\kappa < \kappa_0$ the map (10.4) and

$$\bar{P}: \begin{pmatrix} u \\ v \\ w \end{pmatrix} \rightarrow \begin{pmatrix} Au \\ Bv \\ \omega(w) + R_3(g_2(w), g_1(g_2(w),w), w) \end{pmatrix} \tag{10.5}$$

are topologically equivalent. Here $v = g_1(u,w)$ represents the unique global center - stable manifold for (10.4) and $u = g_2(w)$ represents the unique global center manifold for the map

$$Q: \begin{pmatrix} u \\ w \end{pmatrix} \rightarrow \begin{pmatrix} Au + R_1(u, g_1(u,w), w) \\ \omega(w) + R_3(u, g_1(u,w), w) \end{pmatrix}.$$

<u>Proof of Theorem 10.1</u>: For the purpose of this proof we may identify the given system (10.1) with the transformed system (10.3). First, we have to show how Theorem 10.2 may be applied to system (10.3). Since this goes along the lines of Section 7 a sketch should suffice. By assumption (iv) the spectral radii of A^- and $(A^+)^{-1}$ are less than 1 whereas the spectral radius of A^o equals 1. Choosing a suitable norm on $\mathbb{R}^n = \mathbb{R}^{n_-} \oplus \mathbb{R}^{n_+} \oplus \mathbb{R}^{n_o}$ we manage that all hypotheses of Theorem 10.2 are satisfied for a map which coincides with the map

$$\tilde{P}: \begin{pmatrix} u \\ v \\ w \end{pmatrix} \longrightarrow \begin{pmatrix} A^-u + r_1(u,v,w) \\ A^+v + r_2(u,v,w) \\ A^ow + r_3(u,v,w) \end{pmatrix}$$

which is associated with the difference equation (10.3) on a neighborhood B_σ of $0 \in \mathbb{R}^n$. Thus, by Theorem 10.2 there is a homeomorphism H which maps the solutions of (10.1) onto the solutions of

$$\begin{aligned} u(k+1) &= A^-u(k) \\ v(k+1) &= A^+v(k) \\ w(k+1) &= A^ow(k) + r_3(s_1(w(k)),s_2(w(k)),w(k)) \end{aligned} \tag{10.6}$$

in the neighborhood $H(B_\sigma)$ of the coordinate origin. Here

$$(s_1(w),s_2(w)) := (g_2(w),g_1(g_2(w),w))$$

is the unique global center manifold for the map (10.4). Now the properties (i) through (v) of Theorem 10.1 take on the following form. There exists a sequence (not necessarily a solution of (10.6) yet)

$$(u(k),v(k),w(k)) := H(x(k)) \tag{10.7}$$

such that

(i') there exists a sequence $k_\nu \in \mathbb{N}_0$, $k_\nu \to \infty$ as $\nu \to \infty$, with

$$\lim_{\nu \to \infty} (u(k_\nu), v(k_\nu), w(k_\nu)) = (0,0,0), \tag{10.8}$$

(ii') the zero solution of

$$w(k+1) = A^o w(k) + r_3(s_1(w(k)), s_2(w(k)), w(k)) \tag{10.9}$$

is stable both for $k \to \infty$ and $k \to -\infty$,

(iii') there exists an $\mathbb{R}^n$-ball B_σ $(\subset H(B_\rho))$ around $(u,v,w) = (0,0,0)$ with radius $\sigma > 0$ such that for any ω-limit point $(u_\infty, v_\infty, w_\infty)$ of (10.7) one has

$$u_\infty = 0, \quad v_\infty = 0, \tag{10.10}$$

(iv') $$(u(k), v(k), w(k)) \text{ is bounded as } k \to \infty, \tag{10.11}$$

(v') $$\lim_{k \to \infty} [(u(k+1), v(k+1), w(k+1)) - (u(k), v(k), w(k))] = 0. \tag{10.12}$$

Furthermore, since H carries solutions of (10.1) near $x = 0$ onto solutions of (10.6) near $(u,v,w) = (0,0,0)$ and because of (i') and (v') there exists for each ν sufficiently large (w.l.o.g. for each $\nu \geq 0$) a maximal set $J_\nu := \{\kappa_\nu, \ldots, K_\nu\}$ of consecutive integers with $-\infty \leq \kappa_\nu < k_\nu < K_\nu \leq \infty$ such that

$$(u(k), v(k), w(k)) \text{ is a solution of (10.6) with } (u(k), v(k), w(k)) \in B_\sigma \text{ on } J_\nu \tag{10.13}$$

and

$$\frac{\sigma}{2} \leqq \| (u(\kappa_\nu),v(\kappa_\nu),w(\kappa_\nu)) \| \leqq \sigma \text{ for all } \nu \in \mathbb{N}_0,$$

$$\frac{\sigma}{2} \leqq \| (u(K_\nu),v(K_\nu),w(K_\nu)) \| \leqq \sigma \text{ for all } \nu \in \mathbb{N}_0. \tag{10.14}$$

Our final goal in this proof is to show that

$$\lim_{k\to\infty} (u(k),v(k),w(k)) = (0,0,0). \tag{10.15}$$

We distinguish two cases.

Case 1: For some $\nu = \nu_0$ the set J_ν is unbounded to the right. Then the function (10.7) is a solution of system (10.6) for $k \geqq \kappa_{\nu_0}$ and this immediately implies that

$$u(k) = (A^-)^{k-\kappa_{\nu_0}} u(\kappa_{\nu_0}) \text{ for all } k \geqq \kappa_{\nu_0},$$

$$v(k) = (A^+)^{k-\kappa_{\nu_0}} v(\kappa_{\nu_0}) \text{ for all } k \geqq \kappa_{\nu_0}.$$

The spectral assumptions on A^- and A^+ imply then that

$$\lim_{k\to\infty} u(k) = 0.$$

Moreover, $v(k)$ tends to infinity as k goes to infinity unless $v(\kappa_{\nu_0}) = 0$. Because of (10.13), however, $v(k)$ is bounded on $J_{\nu_0} = \{\kappa_{\nu_0}, \kappa_{\nu_0}+1, \ldots.\}$ and thus $v(\kappa_{\nu_0}) = 0$, i.e.

$$v(k) = 0 \text{ for all } k \geqq \kappa_{\nu_0}.$$

In order to prove the relation $\lim_{k\to\infty} w(k) = 0$ we suppose the contrary, i.e. we suppose that there exists a positive ε and a subsequence $\{k_{\nu_\mu}\}$ of $\{k_\nu\}$ with $k_{\nu_\mu} \to \infty$ as $\mu \to \infty$ such that

$$\| w(k_{\nu_\mu}) \| \geq \varepsilon > 0 \text{ for all } \mu \in \mathbb{N}_0. \tag{10.16}$$

Let $\delta = \delta(\varepsilon)$ be the modulus of stability of the trivial solution of system (10.9) (understood as for differential equations). Because of relation (10.8) there exists a $k_{\bar{\nu}}$ such that

$$\| w(k_{\bar{\nu}}) \| < \delta$$

and therefore

$$\| w(k) \| < \varepsilon \quad \text{for all } k \geq k_{\bar{\nu}}.$$

This, however, contradicts (10.16) and herewith proves (10.15) in case 1.

Case 2: For each ν the set J_ν is bounded to the right which means that the sequence (10.7) leaves the ball B_σ again and again. The estimate (10.13) implies that there exists a convergent subsequence $(u(K_{\nu_\mu}), v(K_{\nu_\mu}), w(K_{\nu_\mu}))$ of $(u(K_\nu), v(K_\nu), w(K_\nu))$ whose limit $(u_\infty, v_\infty, w_\infty)$, say, is then an ω - limit point of (10.7). Because of (iii'), this means that both u_∞ and v_∞ vanish and, with (10.14), that the inequality

$$\frac{\sigma}{2} \leq \| w_\infty \| \leq \sigma$$

holds true. Since the sequence $w(K_{\nu_\mu})$ converges to w_∞ there exists a μ_o such that

$$\| w(k_{\nu_\mu}) \| \geq \frac{\sigma}{4} \quad \text{for all } \mu \geq \mu_o. \tag{10.17}$$

With $\delta = \delta(\sigma/4)$ being the modulus of stability of the zero solution of (10.9) we find a $\bar{\mu}$ such that

$$\| w(k_{\nu_{\bar{\mu}}}) \| < \delta. \tag{10.18}$$

Thus, we get the estimate

$$\| w(k) \| < \frac{\sigma}{4} \text{ for all } k \geq k_{\nu_{\bar{\mu}}}$$

which contradicts (10.17). This shows that case 2 does not occur and so Theorem 10.1 is proved. ∎

11. The Basic Selection Model from Population Genetics for Separated Generations

In this section we investigate the convergence problem in the Fisher-Wright-Haldane model for separated generations. Except for the discrete rather than the continuous time the setting is the same as in Section 6. $a_1,\ldots,a_n$ are the n *alleles* with corresponding *gene frequencies* x_i, $i = 1,\ldots,n$. α_{jk} is the relative frequency of the *genotype* (a_j,a_k), $j,k = 1,\ldots,n$. Again, $F = (f_{jk})$ is the symmetric, nonnegative matrix of *viability parameters* and $\Phi(x) = x^T Fx$ is the *mean viability* of the population at state x.

The difference equation for the evolution of the gene frequency vector $x = (x_1,\ldots,x_n)^T$ can be derived rather easily. The genotype frequencies α_{jk} after one generation become

$$f_{jk}\alpha_{jk} / \sum_{r,s=1}^{n} f_{rs}\alpha_{rs}, \quad j,k = 1,\ldots,n,$$

and so the gene frequencies x_j (recall the *random mating* assumption $\alpha_{jk} = x_j x_k$) after selection are

$$x_j [\sum_{k=1}^{n} f_{jk}x_k] / \sum_{r,s=1}^{n} f_{rs}x_r x_s, \quad j = 1,\ldots,n.$$

Denoting by $x(k)$ the gene frequency vector at generation k, $k \in \mathbb{N}_o$, the difference equation for $x(k)$ may be written as

$$x(k+1) = H(x(k)) \qquad (11.1)$$

where

$$H(x) := \frac{1}{\Phi(x)} \begin{pmatrix} x_1 & & 0 \\ & \ddots & \\ 0 & & x_n \end{pmatrix} F \begin{pmatrix} x_1 \\ \vdots \\ x_n \end{pmatrix}.$$

At those states where the mean viability Φ vanishes the population dies out after one generation. So the appropriate state space is the set

$$\Lambda_o^{n-1} := \{x \in \Lambda^{n-1}: \Phi(x) > 0\} \text{ (see (6.5))}$$
$$= \{x \in \mathbb{R}^n: \sum_{i=1}^{n} x_i = 1,\ x_i \geq 0,\ i = 1,\dots,n,\ \Phi(x) > 0\}$$

whose invariance will be shown in a moment. For any $x_o \in \Lambda_o^{n-1}$ we define the corresponding solution of (11.1) for $k \geq 0$ by

$$x(k,x_o) := H^k(x_o),$$

the k - th iterate of x_o under the map H. The ω - limit set, i.e. the set of accumulation points, of $x(k,x_o)$ is denoted by $\Omega(x_o)$ and the set of stationary solutions of (11.1) in Λ_o^{n-1} is

$$E_o^{n-1} := \{x \in \Lambda_o^{n-1}: \operatorname{diag}(x)Fx = \Phi(x)x\}.$$

The invariance of Λ_o^{n-1} under (11.1) can be seen as follows. First, for any $x \in \Lambda_o^{n-1}$ one gets $e^T H(x) = \Phi(x)^{-1} e^T \operatorname{diag}(x)Fx = \Phi(x)^{-1}\Phi(x) = 1$ and so any solution starting in Λ_o^{n-1} remains in the hyperspace $e^T x = 1$. Hence, in order to show that Λ_o^{n-1} is positively invariant it suffices to prove that solutions starting in Λ_o^{n-1} remain in

$$\mathbb{R}^n_{+,o} := \{x \in \mathbb{R}^n : x_i \geq 0, \ i = 1,\ldots,n, \ \Phi(x) > 0\}. \qquad (11.2)$$

In order to show this we take any $x_o \in \Lambda_o^{n-1}$ and assume without loss of generality that x_o has the form

$$x_o = \begin{pmatrix} \tilde{x}_o \\ 0 \end{pmatrix}, \quad \tilde{x}_o > 0$$

and partition F accordingly

$$F = \begin{pmatrix} F_{11} & F_{12} \\ F_{21} & F_{22} \end{pmatrix}.$$

Suppose that $\Phi(x_o)$ is positive, then F_{11} is different from the zero matrix. This implies that

$$H(x_o) = \Phi(x_o)^{-1} \begin{pmatrix} \operatorname{diag}(\tilde{x}_o)F_{11}\tilde{x}_o \\ 0 \end{pmatrix},$$

and since $\operatorname{diag}(\tilde{x}_o)F_{11}\tilde{x}_o$ is a nonnegative vector, $H(x_o)$ lies in the nonnegative orthant $\mathbb{R}^n_+$ of $\mathbb{R}^n$. The inequality $\Phi(H(x_o)) > 0$, however, follows from the positiveness of

$$\Phi(H(x_o)) = H(x_o)^T F H(x_o) = \Phi(x_o)^{-2}\tilde{x}_o^T F_{11}\operatorname{diag}(\tilde{x}_o)F_{11}\operatorname{diag}(\tilde{x}_o)F_{11}\tilde{x}_o$$

which in turn is a consequence of the fact that $\operatorname{diag}(\tilde{x}_o)F_{11}\tilde{x}_o$ is a nonnegative vector different from the zero vector. This proves the assertion that Λ_o^{n-1} is positively invariant under the flow of the difference equation (11.1).

Now we are prepared to attack the problem of this section. We want to investigate the asymptotic behavior of the solutions of (11.1). In the three - dimensional case, i.e. the case with a two - dimensional state space, it has been shown by Feller [15] and an der Heiden [25] that any solution of (11.1) converges to a stationary solution. Our answer to this question in the general case is given in Theorem 11.2 below.

As in the continuous time case the first step in the analysis of the convergence problem can be done by exploiting LaSalle's Theorem which also exists for discrete time systems.

Theorem 11.1 (LaSalle [32, Chapter 2, Theorem 6.4]): Let K be any subset of Λ_o^{n-1} and suppose that there exists a function $V \in C(\mathbb{R}^n_{+,o}, \mathbb{R})$ (see (11.2)) such that the function

$$\dot{V}(x) := V(H(x)) - V(x)$$

is nonnegative on K.

Then, for any $x_o \in K$ with the property that $x(k,x_o)$ is in K for all $k \geq 0$, one has

$$\Omega(x_o) \subset \{x \in \bar{K}: \dot{V}(x) = 0\}.$$

The proof that the mean viability $\Phi(x)$ is a suitable Ljapunov function for (11.1) requires an inequality which is well known in population genetics. We state this result and refer for the proof e.g. to Hadeler [20, p.99].

Lemma 11.1: For all $x \in \Lambda_o^{n-1}$ the following holds true:

(i) $$x^T F \operatorname{diag}(x) F \operatorname{diag}(x) Fx \geq \Phi(x)^3,$$

(ii) equality in (i) holds if and only if

$$\operatorname{diag}(x)\, F x = \Phi(x) x.$$

Using this lemma together with LaSalle's Theorem we get the next auxiliary result.

<u>Lemma 11.2</u>: For any $x_o \in \Lambda_o^{n-1}$ the ω - limit set $\Omega(x_o)$ is contained in the set E_o^{n-1} of stationary solutions of (11.1).

<u>Remark</u>: The ω - limit set of a solution of a discrete dynamical system need not be connected in general. In our case, however, $\Omega(x_o)$ is connected for any $x_o \in \Lambda_o^{n-1}$ since it contains only stationary solutions of (11.1). For a proof of this see e.g. Hadeler [20, p.102]. Therefore, as in the continuous time case, the convergence problem is settled if E_o^{n-1} is known to consist of isolated points. Otherwise, i.e. in the general case, a further analysis in the spirit of this book is necessary.

<u>Proof of Lemma 11.2</u>: We apply LaSalle's Theorem with $K := \Lambda_o^{n-1}$ and $V := \Phi$. The derivative $\dot{\Phi}$ of Φ with respect to (11.1) has the form

$$\dot{\Phi}(x) = \Phi(x)^{-2} x^T \operatorname{diag}(x)\, F \operatorname{diag}(x)\, F x - \Phi(x) \qquad (11.3)$$

and from Lemma 11.1 it follows that $\dot{\Phi}$ does not change sign on Λ_o^{n-1}. Since Λ_o^{n-1} is bounded and positively invariant, LaSalle's Theorem says that for any $x_o \in \Lambda_o^{n-1}$

$$\Omega(x_o) \subset \{x \in \text{closure } \Lambda_o^{n-1} : \dot{\Phi}(x) = 0\}. \qquad (11.4)$$

From Lemma 11.1 and the relation (11.3) it follows that the equality $\dot{\Phi}(x) = 0$ implies that $\mathrm{diag}(x)\,F\,x = \Phi(x)x$, thus the set on the right hand side of (11.4) is contained in E^{n-1} (see Section 6). In order to prove the inclusion $\Omega(x_o) \subset E_o^{n-1}$ it remains to show that $\Phi(y)$ is positive for any $y \in \Omega(x_o)$. The condition $y \in \Omega(x_o)$, however, means that there exists a sequence of integers $k' \to \infty$ such that

$$y = \lim_{k' \to \infty} H^{k'}(x_o) .$$

Since $\dot{\Phi}$ is nonnegative on Λ^{n-1} (see Section 6) the inequality

$$\Phi(H^{k'}(x_o)) - \Phi(x_o) \geq 0 \text{ for all } k'$$

is valid and because $\Phi(x_o)$ is positive this implies that

$$\Phi(y) = \lim_{k' \to \infty} \Phi(H^{k'}(x_o)) > 0.$$

Thus, the proof of Lemma 11.2 is complete. ∎

In the formulation of the remaining results of this section we make use of the notions *index*, *defect*, *regular point* and *exceptional point* which have been introduced in Section 6.

> Lemma 11.3: Let $\bar{x}$ be a regular stationary solution of (11.1) and let $d = d(\bar{x})$ be its defect.
> Then there exists an $\mathbb{R}^n$- neighborhood U of $\bar{x}$ such that $E_o^{n-1} \cap U$ is a smooth manifold of dimension d.

Proof: Given a stationary solution $\bar{x} \in E_o^{n-1}$, particularly meaning that

$\Phi(\bar{x})$ is positive, there exists a neighborhood $\bar{U}$ of $\bar{x}$ such that Φ does not vanish in $\bar{U}$. Hence, we get the relation

$$E_o^{n-1} \cap \bar{U} = E^{n-1} \cap \bar{U}.$$

As in Lemma 6.3 we may then apply Theorem C.2, Appendix C to complete the proof. ■

The last lemma we need for the proof of the main theorem of this section is the discrete analogue of Lemma 6.4.

Lemma 11.4: Let $\bar{x}$ be a regular point of (11.1) and let $d = d(\bar{x})$ be its defect.
Then the Jacobian $H_x(\bar{x})$ of H at $\bar{x}$ has $n-d$ eigenvalues with moduli different from 1.

Proof: Again we rearrange the state vector components so that $\bar{x}$ and F, respectively, appear in the form

$$\bar{x} = \begin{pmatrix} \tilde{x} \\ 0 \end{pmatrix}, \quad F = \begin{pmatrix} F_j & F_{12} \\ F_{21} & F_{22} \end{pmatrix}, \tag{11.5}$$

where $\tilde{x} \in \mathbb{R}^{n_j}$, $\tilde{x} > 0$, $F_j \in \mathbb{R}^{n_j \times n_j}$. Here $j = j(\bar{x})$ is the index of $\bar{x}$. The regularity assumption of $\bar{x}$ means that

none of the components of the vector $F_{21}\tilde{x}$ equals $\Phi(\bar{x})$. (11.6)

In order to find a suitable expression for the Jacobian $H_x(\bar{x})$ we expand

H with respect to $y = x - \bar{x}$ around $y = 0$.

$$\begin{aligned}
H(y) &= \Phi(\bar{x}+y)^{-1}[\mathrm{diag}(\bar{x}+y)F(\bar{x}+y)] = \\
&[\Phi(\bar{x}) + 2\bar{x}^T F y]^{-1}[\mathrm{diag}(\bar{x})F\bar{x} + \mathrm{diag}(\bar{x})Fy + \mathrm{diag}(y)F\bar{x}] + o(y^T y) = \\
&\bar{x} + \Phi(\bar{x})^{-1}[\mathrm{diag}(\bar{x})Fy + \mathrm{diag}(y)F\bar{x} - 2(\bar{x}^T F y)\bar{x}] + o(y^T y) = \\
&\bar{x} + \Phi(\bar{x})^{-1}[\mathrm{diag}(\bar{x})F + \mathrm{diag}(F\bar{x}) - 2\bar{x}\bar{x}^T F]y + o(y^T y).
\end{aligned}$$

Taking into account the particular form (11.5) of $\bar{x}$ and F, the Jacobian $H_x(\bar{x})$ assumes the form

$$H_x(\bar{x}) = \begin{pmatrix} H_{11}(\bar{x}) & H_{12}(\bar{x}) \\ 0 & H_{22}(\bar{x}) \end{pmatrix}, \tag{11.7}$$

where

$$\begin{aligned}
H_{11}(\bar{x}) &:= \Phi(\bar{x})^{-1}[\mathrm{diag}(\tilde{x})F_j + \mathrm{diag}(F_j\tilde{x}) - 2\tilde{x}\tilde{x}^T F_j], \\
H_{12}(\bar{x}) &:= \Phi(\bar{x})^{-1}[\mathrm{diag}(\tilde{x}) - 2\tilde{x}\tilde{x}^T]F_{12}, \\
H_{22}(\bar{x}) &:= \Phi(\bar{x})^{-1}\mathrm{diag}(F_{21}\tilde{x}).
\end{aligned}$$

$\bar{x}$ is a stationary solution of (11.1) and $\tilde{x}$ is a positive vector. Thus, $\tilde{x}$ satisfies the relation

$$F_j\tilde{x} = \Phi(\bar{x})e^j \tag{11.8}$$

(note that $e^j = (1,\dots,1)^T \in \mathbb{R}^{n_j}$) and with this the expression for $H_{11}(\bar{x})$ simplifies to

$$H_{11}(\bar{x}) = \Phi(\bar{x})^{-1}\mathrm{diag}(\tilde{x})F_j + I_{n_j} - 2\tilde{x}(e^j)^T. \tag{11.9}$$

The regularity assumption (11.6) implies that the lower right block $H_{22}(\bar{x})$ of $H_x(\bar{x})$ is a real diagonal matrix which does not have the eigenvalue 1. Thus, in order to prove the lemma it suffices to prove the claim: $H_{11}(\bar{x})$ has $n_j - d$ eigenvalues with moduli $\neq 1$. (11.10)

In contrast to the continuous time situation here we have to deal only with the case $\Phi(\bar{x}) > 0$. This in particular means that $F_j \neq 0_{n_j \times n_j}$. Recalling the abbreviation

$$\tilde{X} = \operatorname{diag}((\tilde{x}_1)^{1/2}, \ldots, (\tilde{x}_{n_j})^{1/2})$$

we get for the matrix

$$\Phi(\bar{x})^{-1}\tilde{X}F_j\tilde{X} \tag{11.11}$$

the eigenvalue - eigenvector relation

$$(\Phi(\bar{x})^{-1}\tilde{X}F_j\tilde{X})\tilde{X}^{-1}\tilde{x} = \Phi(\bar{x})^{-1}\tilde{X}\Phi(\bar{x})e^j = \tilde{X}e^j = \tilde{X}^{-1}\tilde{x} \tag{11.12}$$

which says that the matrix (11.11) has the eigenvalue 1 with corresponding eigenvector $\tilde{X}^{-1}\tilde{x}$. The matrix (11.11) is related to F_j via Sylvester's Inertia Theorem. By assumption, the symmetric matrix F_j has a d-dimensional kernel $(d < n_j)$ and ipso facto $n_j - d$ nonzero eigenvalues. Then also the matrix (11.11) has $n_j - d$ nonzero eigenvalues $\lambda_1, \ldots, \lambda_{n_j - d}$, where we take for definiteness λ_1 as the eigenvalue 1. The symmetry of $\Phi(\bar{x})^{-1}\tilde{X}F_j\tilde{X}$ allows us to choose an orthogonal set $\{\tilde{y}_1, \ldots, \tilde{y}_{n_j - d}\}$ of eigenvectors associated with these eigenvalues, i.e.

$$(\Phi(\bar{x})^{-1}\tilde{X}F_j\tilde{X})\tilde{y}_i = \lambda_i\tilde{y}_i \text{ for } i = 1, \ldots, n_j - d \tag{11.13}$$

where $\tilde{y}_1 := \tilde{X}^{-1}\tilde{x}$ (see (11.12)). We shall particularly use the orthogonality relations

$$(\tilde{y}_1)^T\tilde{y}_i = \tilde{x}^T\tilde{X}^{-1}\tilde{y}_i = 0 \text{ for } i = 1, \ldots, n_j - d$$

or equivalently,

$$(e^j)^T\tilde{X}\tilde{y}_i = 0 \text{ for } i = 2,\dots,n_j-d. \tag{11.14}$$

Next we derive n_j-d eigenvalue - eigenvector relations for the matrix $H_{11}(\bar{x})$. First of all we get with (11.9) and (11.12)

$$H_{11}(\bar{x})\tilde{x} = \tilde{X}(\Phi(\bar{x})^{-1}\tilde{X}F_j)\tilde{x} + \tilde{x} - 2\tilde{x}(e^j)^T\tilde{x} = \tilde{X}\tilde{X}^{-1}\tilde{x} + \tilde{x} - 2\tilde{x} = 0,$$

i.e. $H_{11}(\bar{x})$ admits the eigenvalue 0. Note here, that because of this eigenvalue 0 the map H cannot be a diffeomeorphism near any regular point. Next, with (11.13) and (11.14) we get for $i = 2,\dots,n_j-d$

$$H_{11}(\bar{x})\tilde{X}\tilde{y}_i = \tilde{X}(\Phi(\bar{x})^{-1}\tilde{X}F_j)\tilde{X}\tilde{y}_i + \tilde{X}\tilde{y}_i - 2\tilde{x}(e^j)^T\tilde{X}\tilde{y}_i = (\lambda_i + 1)\tilde{X}\tilde{y}_i.$$

Hence, $H_{11}(\bar{x})$ has the n_j-d real eigenvalues $0,\lambda_2+1,\dots,\lambda_{n_j-d}+1$. In order to show that all these eigenvalues have moduli different from 1 it remains to be shown that $\lambda_i \neq -2$ for $i = 2,\dots,n_j-d$. This, however, is true because not only the eigenvalues $\lambda_2,\dots,\lambda_{n_j-d}$ of $\Phi(\bar{x})^{-1}\tilde{X}F_j\tilde{X}$ but all eigenvalues of this matrix lie in the compact interval [-1,1]. To see this we consider the spectral radius of the matrix

$$\frac{1}{\Phi(\bar{x})}\tilde{X}^2F_j = \frac{1}{\Phi(\bar{x})}\operatorname{diag}(\tilde{x})F_j,$$

which is similar to $\Phi(\bar{x})^{-1}\tilde{X}F_j\tilde{X}$ and therefore both matrices have the same spectral radius. The equality (use (11.8))

$$(e^j)^T\Phi(\bar{x})^{-1}\operatorname{diag}(\tilde{x})F_j = \Phi(\bar{x})^{-1}\tilde{x}^TF_j = \Phi(\bar{x})^{-1}\Phi(\bar{x})(e^j)^T = (e^j)^T$$

shows that each row sum of the matrix $\Phi(\bar{x})^{-1}\operatorname{diag}(\tilde{x})F_j$ equals 1. The elements of this matrix are nonnegative and so its row sum norm is 1.

Consequently (see e.g. Stoer and Bulirsch [39]) the spectral radius is not greater than 1. This completes the proof of the lemma. ∎

The main result of this section is as follows.

Theorem 11.2: The ω - limit set $\Omega(x_o)$ of any solution $x(k,x_o)$ of equation (11.1) on Λ_o^{n-1} is either a continuum of exceptional points or a single point.

Proof: As mentioned in the remark succeeding Lemma 11.2 the ω - limit set $\Omega(x_o)$ is connected. Since $\Omega(x_o)$ consists of stationary solutions of (11.1) it is a continuum of exceptional points or it contains at least one regular point. Theorem 11.2 is proved when it is shown that any regular ω - limit point of the solution $x(k,x_o)$ is the limit of this solution as $k \to \infty$.

So let $\bar{x}$ be a regular point in $\Omega(x_o)$. Thus, $\Omega(x_o)$ is not empty and hypothesis (i) of Theorem 7.1 is satisfied. Lemma 11.2 says that $\Omega(x_o)$ is contained in E_o^{n-1} and by Lemma 11.3 there exists a neighborhood U of $\bar{x}$ such that $\Omega(x_o) \cap U$ is contained in the smooth d - dimensional manifold $E_o^{n-1} \cap U$ of stationary solutions of (11.1). Hence, hypothesis (ii) of Theorem 7.1 is fulfilled and with Lemma 11.4 condition (iii) as well. Since the state space Λ_o^{n-1} of system (11.1) is bounded also the solution $x(k,x_o)$ is bounded. Finally, as mentioned above, the validity of (v) follows from the fact that any ω - limit point of $x(k,x_o)$ is a stationary solution of (11.1). Thus, Theorem 7.1 applies and Theorem 11.2 is proved. ∎

Theorem 11.2 shows that all results which have been proved in Aulbach and Hadeler [7] for the continuous time case are equally true in the discrete time case. As in Section 6 we finally get a general convergence result if the underlying dynamical system is supposed to be *cooperative*, i.e. if any solution of (11.1) in the relative interior of Λ_o^{n-1} stays away in the future from the relative boundary of Λ_o^{n-1}.

<u>Corollary</u>: If system (11.1) is cooperative, then any solution in the relative interior int Λ_o^{n-1} of Λ_o^{n-1} converges to a stationary solution in int Λ_o^{n-1}.

APPENDICES

Appendix A: Reducibility

For the transformation of system (2.2) into the form (2.4) we need a reducibility result which generalizes a theorem of Coppel [12, Chapter 5]. In order to prove our result we have to have two lemmas which can be found, more or less explicitly stated, in Coppel [12, Chapter 5] and Harris and Miles [23, Appendix 2]. In either case those results are stated and proved in a complex setting whereas in this book we deal with real systems. Since it is not obvious from the outset that during the proof one may stay within the field of real numbers we give a proof for either lemma.

Lemma A.1: Let P be an orthogonal projection of $\mathbb{R}^n$, i.e. $P^2 = P = P^T$. Let $\Phi(t)$ be a real nonsingular $n\times n$ matrix which depends on a nonnegative parameter t.
Then, for each $t \geq 0$, there exists a real nonsingular matrix $S(t)$ such that for each $t \geq 0$

(i) $S(t)PS^{-1}(t) = \Phi(t)P\Phi^{-1}(t)$,

(ii) $\| S(t) \|_2 \leq 2^{1/2}$,

(iii) $\| S^{-1}(t) \|_2 \leq [\| \Phi(t)P\Phi^{-1}(t) \|_2^2 + \| \Phi(t)(I_n - P)\Phi^{-1}(t) \|_2^2]^{1/2}$.

If $\Phi(t)$ is continuously differentiable on the interval $[0,\infty)$, then

(iv) $S(t)$ is continuously differentiable on $[0,\infty)$.

Proof: Throughout this proof we suppress the argument t. It is straightforward to verify that the matrix

$$P\Phi^T\Phi P + (I_n - P)\Phi^T\Phi(I_n - P) \tag{A.1}$$

is symmetric and positive definite. First we show that there exists an $n\times n$ matrix

$$R\text{, real, symmetric and positive definite} \tag{A.2}$$

such that

$$R^2 = P\Phi^T\Phi P + (I_n - P)\Phi^T\Phi(I_n - P). \tag{A.3}$$

It is well known that any positive definite Hermitian matrix has a unique positive definite square root, so we have to show that in our case the square root of (A.1) is real.

There exists an orthogonal matrix M such that (A.1) equals M^TDM where $D = \text{diag}(d_1,\dots,d_n)$ is a diagonal matrix with positive elements $d_1,\dots,d_n$ along the main diagonal. Defining

$$\bar{D} := \text{diag}(d_1^{1/2},\dots,d_n^{1/2}) \tag{A.4}$$

it is clear that the real matrix

$$R := M^T\bar{D}M \tag{A.5}$$

meets our purposes and (A.2) and (A.3) are proved.

Next we show that R commutes with P, i.e. that

$$RP = PR. \tag{A.6}$$

In order to see this we first realize the relation

$$R^2P = PR^2 \tag{A.7}$$

which follows immediately from (A.3). Using the representation (A.4), (A.5) of R and denoting the elements of the matrix MPM^T by m_{ij}, $i,j = 1,\dots,n$, the assertion (A.6) reads

$$d_i^{1/2} m_{ij} = d_j^{1/2} m_{ij}, \quad i,j = 1,\dots,n,$$

whereas the known relation (A.7) appears as

$$d_i m_{ij} = d_j m_{ij}, \quad i,j = 1,\dots,n.$$

Distinguishing between m_{ij} being zero or nonzero the remainder of the proof of (A.6) is obvious.

With the definition

$$S := \Phi R^{-1} \tag{A.8}$$

the validity of the assertion (i) in Lemma A.1 follows readily from (A.6). In order to prove (ii) we notice the relation

$$PS^TSP + (I_n - P)S^TS(I_n - P) = I_n$$

whose verification is straightforward, using (A.3) and the commutativity of $(I_n - P)$ and $(R^{-1})^T$. With this we get for all $x \in \mathbb{R}^n$

$$\| Sx\|_2^2 = \| SPx + S(I_n - P)x\|_2^2 \leq [\| SPx\|_2 + \| S(I_n - P)x\|_2]^2 \leq$$

$$2[\| SPx\|_2^2 + \| S(I_n - P)x\|_2^2] =$$

$$2[x^T PS^T SPx + x^T(I_n - P)S^T S(I_n - P)x] =$$

$$2x^T I_n x = 2\| x\|_2^2 ,$$

where $\| \cdot \|_2$ denotes the Euclidean norm. Thus, the estimate

$$\| S\|_2 \leq 2^{1/2}$$

is proved. For the proof of (iii) we conclude from (A.3)

$$(S^{-1})^T S^{-1} = (\Phi^T)^{-1} P\Phi^T \Phi P\Phi^{-1} + (\Phi^T)^{-1}(I_n - P)\Phi^T \Phi(I_n - P)\Phi^{-1}$$

and from this, for all $x \in \mathbb{R}^n$,

$$\| S^{-1}x\|_2^2 = \| \Phi P\Phi^{-1}x\|_2^2 + \| \Phi(I_n - P)\Phi^{-1}x\|_2^2 \leq$$

$$[\| \Phi P\Phi^{-1}\|_2^2 + \| \Phi(I_n - P)\Phi^{-1}\|_2^2] \| x\|_2^2 .$$

This proves (iii). Finally it follows from a result of Reid [38, Lemma 3.1] that a continuously differentiable Φ leads to a continuously differentiable S. ■

Lemma A.2: Let P be an orthogonal projection of $\mathbb{R}^n$ and let $\Phi(t)$ be a fundamental matrix of the linear system

$$\dot{x} = A(t)x, \qquad (A.9)$$

where A(t) is continuous and bounded on $[0,\infty)$. Then, if

$$\| \Phi(t)P\Phi^{-1}(t) \| \text{ is bounded on } [0,\infty),$$

the matrix $S(t)$ associated with $\Phi(t)$ by Lemma A.1 has a bounded derivative $\dot{S}(t)$ on $[0,\infty)$.

Proof: Since the assumptions of Lemma A.2 on P and $\Phi(t)$ are stronger than those of Lemma A.1 we may liberally use all relations which have been derived throughout the proof of Lemma A.1. Let $S(t)$ and $R(t)$ be the continuously differentiable matrices which have been defined there. The change of variables

$$x = S(t)y$$

transforms system (A.9) into

$$\dot{y} = B(t)y, \quad t \geq 0, \tag{A.10}$$

where

$$B(t) := S^{-1}(t)[A(t)S(t) - \dot{S}(t)]. \tag{A.11}$$

By (A.8) the matrix $R(t)$ is a fundamental matrix of system (A.10) and so we get the identity

$$B(t) = \dot{R}(t)R^{-1}(t). \tag{A.12}$$

Since by assumption $A(t)$ is bounded on $[0,\infty)$ there exists a positive constant θ such that

$$-\theta I_n \leq A(t) + A^T(t) \leq \theta I_n \text{ for all } t \geq 0, \tag{A.13}$$

where the relation $B \leq C$ for matrices B and C means that $C - B$ is positive semidefinite. Differentiation of (A.3) gives

$$R(t)\dot{R}(t) + \dot{R}(t)R(t) = P\Phi^T(t)[A(t) + A^T(t)]\Phi(t)P +$$
$$(I_n - P)\Phi^T(t)[A(t) + A^T(t)]\Phi(t)(I_n - P)$$

and with (A.13) this implies that the matrix

$$R(t)\dot{R}(t) + \dot{R}(t)R(t) + \theta R^2(t) =$$
$$P\Phi^T(t)[A(t) + A^T(t) + \theta I_n]\Phi(t)P +$$
$$(I_n - P)\Phi^T(t)[A(t) + A^T(t) + \theta I_n]\Phi(t)(I_n - P)$$

is positive semidefinite. Using the same argument for $A(t) + A^T(t) - \theta I_n$ (see (A.13)) we conclude that

$$-\theta R^2(t) \leq R(t)\dot{R}(t) + \dot{R}(t)R(t) \leq \theta R^2(t) \qquad (A.14)$$

and herewith

$$-\theta I_n \leq \dot{R}(t)R^{-1}(t) + R^{-1}(t)\dot{R}(t) \leq \theta I_n \text{ for all } t \geq 0.$$

From this we obtain

$$[\dot{R}(t)R^{-1}(t) + R^{-1}(t)\dot{R}(t)]^2 \leq \theta^2 I_n$$

and further

$$\| \dot{R}(t)R^{-1}(t) + R^{-1}(t)\dot{R}(t) \|_* \leq n^{1/2}\theta, \qquad (A.15)$$

where $\| K \|_* := (\text{trace}(K^TK))^{1/2}$. With this inequality we now show that (see (A.11)) the matrix

$$\dot{S}(t) = A(t)S(t) - S(t)B(t)$$

is bounded on $[0,\infty)$. Since by Lemma A.1 the matrix $S(t)$ is bounded on $[0,\infty)$ and since $A(t)$ is bounded by assumption the proof of Lemma A.2 is complete when we have shown that $B(t)$ is bounded on $[0,\infty)$.

The trace of a product of matrices is unaltered by cyclic permutation of the factors. Thus, we obtain (abbreviating tr := trace and suppressing the variable t)

$$\begin{aligned} &\operatorname{tr}(\dot{R}R^{-1}+R^{-1}\dot{R})^2 = \operatorname{tr}(\dot{R}R^{-1})^2 + \operatorname{tr}(\dot{R}R^{-1}R^{-1}\dot{R}) + \\ &\operatorname{tr}(R^{-1}\dot{R}^2R^{-1}) + \operatorname{tr}(R^{-1}\dot{R})^2 = 2[\operatorname{tr}(R^{-1}\dot{R}^2R^{-1}) + \operatorname{tr}(R^{-1}\dot{R})^2]. \end{aligned} \tag{A.16}$$

Since the positive semidefinite matrix R^{-1} has a symmetric square root U, say, we may estimate

$$\operatorname{tr}(R^{-1}\dot{R})^2 = \operatorname{tr}(U\dot{R}U)^2 = \| U\dot{R}U \|_*^2 \geq 0$$

and using this and (A.16) we get

$$\begin{aligned} &\| \dot{R}R^{-1} \|_*^2 = \operatorname{tr}(R^{-1}\dot{R}^2R^{-1}) \leq \\ &\frac{1}{2}\operatorname{tr}(\dot{R}R^{-1} + R^{-1}\dot{R})^2 = \frac{1}{2} \| \dot{R}R^{-1} + R^{-1}\dot{R} \|_*^2 . \end{aligned} \tag{A.17}$$

Finally, combining (A.12), (A.17) and (A.15) we arrive at

$$\| B \|_*^2 = \| \dot{R}R^{-1} \|_* \leq \frac{1}{2} \| \dot{R}R^{-1} + R^{-1}\dot{R} \|_*^2 \leq \frac{1}{2} n\theta^2 .$$

Recalling that we have suppressed the t - dependence of the matrices involved in the previous considerations we see that $B(t)$ is bounded on $[0,\infty)$. As mentioned above this proves the boundedness of $\dot{S}(t)$ and completes the proof of Lemma A.2. ∎

Now we are ready to prove the main result of this section which is concerned with linear time varying differential systems of the form

$$\dot{x} = A(t)x, \tag{A.18}$$

where the real $n \times n$ matrix $A(t)$ is supposed to be continuous and bounded on $[0,\infty)$. The crucial assumption is that (A.18) admits an exponential dichotomy which is incomplete in the following sense: there exist projections P^-, P^+ of $\mathbb{R}^n$ and positive constants α, β and γ such that there exists a fundamental matrix $\Phi(t)$ of (A.18) with

$$\| \Phi(t)P^-\Phi^{-1}(s) \| \leq \gamma e^{-\alpha(t-s)} \quad \text{for all } t \geq s \geq 0,$$

$$\| \Phi(t)P^+\Phi^{-1}(s) \| \leq \gamma e^{-\beta(s-t)} \quad \text{for all } s \geq t \geq 0.$$

It should be noted that we allow a strict inequality in

$$n_- + n_+ := \text{rank } P^- + \text{rank } P^+ \leq n$$

whereas in Coppel's Theorem [12, Chapter 5] equality is required, i.e. $P^- + P^+ = I_n$.

Theorem A.1: Under the above hypotheses system (A.18) is *kinematically similar* to a system of the form

$$\begin{aligned} \dot{u} &= A^-(t)u \\ \dot{v} &= A^+(t)v \\ \dot{w} &= A^*(t)w, \end{aligned} \tag{A.19}$$

i.e. there exists a nonsingular real $n \times n$ matrix $T(t)$ which, together with its inverse, is continuously differentiable and bounded on the interval $[0,\infty)$ with bounded derivative such that

$$x = T(t)\begin{pmatrix} u \\ v \\ w \end{pmatrix}$$

transforms system (A.18) into (A.19). System (A.19) has the following properties: the matrices $A^-(t)$, $A^+(t)$, $A^*(t)$ are continuous and bounded on $[0,\infty)$ and the principal fundamental matrices $\Phi^-(t,s)$, $\Phi^+(t,s)$ of $\dot{u} = A^-(t)u$, $\dot{v} = A^+(t)v$, respectively, satisfy estimates of the form

$$\| \Phi^-(t,s) \| \leq \tilde{\gamma} e^{-\alpha(t-s)} \quad \text{for all } t \geq s \geq 0,$$

$$\| \Phi^+(t,s) \| \leq \tilde{\gamma} e^{-\beta(s-t)} \quad \text{for all } s \geq t \geq 0,$$

where $\tilde{\gamma}$ is a positive constant.

Proof: First we can choose a nonsingular $n \times n$ matrix C such that

$$C^{-1}P^-C = \operatorname{diag}(I_{n_-}, O_{n_+}, O_{n-n_--n_+}) =: D^-,$$

$$C^{-1}P^+C = \operatorname{diag}(O_{n_-}, I_{n_+}, O_{n-n_--n_+}) =: D^+.$$

For the fundamental matrix $\Psi(t) := \Phi(t)C$ of (A.18) we get the estimates

$$\begin{aligned} \| \Psi(t)D^-\Psi^{-1}(s) \| &\leq \gamma e^{-\alpha(t-s)} \quad \text{for all } t \geq s \geq 0, \\ \| \Psi(t)D^+\Psi^{-1}(s) \| &\leq \gamma e^{-\beta(s-t)} \quad \text{for all } s \geq t \geq 0. \end{aligned} \tag{A.20}$$

By Lemma A.1 there exists a nonsingular $n \times n$ matrix $S_1(t)$ with

$$D^-S_1^{-1}(t)\Psi(t) = S_1^{-1}(t)\Psi(t)D^- \quad \text{for all } t \geq 0 \tag{A.21}$$

and

$$\|S_1(t)\|_2 \leq 2^{1/2} \text{ on } [0,\infty),$$

$$\|S_1^{-1}(t)\|_2 \leq [\gamma^2 + (\|I_n\|_2 + \gamma)^2]^{1/2} =: K \text{ on } [0,\infty).$$

The transformation $x = S_1(t)y$ carries system (A.18) into

$$\dot{y} = \hat{A}(t)y, \tag{A.22}$$

where

$$\hat{A}(t) := S_1^{-1}(t)[A(t)S_1(t) - \dot{S}_1(t)]$$

is bounded on $[0,\infty)$. Note here that the matrices $S_1(t)$, $S_1^{-1}(t)$ and $\dot{S}_1(t)$ are bounded by Lemmas A.1 and A.2. Since $\hat{\Psi}(t) := S_1^{-1}(t)\Psi(t)$ is a fundamental matrix of (A.22) we get for $\hat{A}(t)$ the representation

$$\hat{A}(t) = \dot{\hat{\Psi}}(t)\hat{\Psi}^{-1}(t)$$

which, in view of (A.21), shows that $\hat{A}(t)$ commutes with D^-. Thus, $\hat{A}(t)$ and ipso facto $\hat{\Psi}(t)$ have block diagonal form, i.e.

$$\hat{A}(t) = \mathrm{diag}(A^-(t), A'(t)),$$

$$\hat{\Psi}(t) = \mathrm{diag}(\Psi^-(t), \Psi'(t)),$$

with n_--dimensional square matrices $A^-(t)$, $\Psi^-(t)$. Because of (A.20) the fundamental matrix $\hat{\Psi}(t)$ of (A.22) satisfies estimates of the form

$$\|\hat{\Psi}(t)D^-\hat{\Psi}^{-1}(s)\| \leq 2^{1/2}K\hat{\gamma}e^{-\alpha(t-s)} \quad \text{for all } t \geq s \geq 0,$$

$$\|\hat{\Psi}(t)D^+\hat{\Psi}^{-1}(s)\| \leq 2^{1/2}K\hat{\gamma}e^{-\beta(s-t)} \quad \text{for all } s \geq t \geq 0,$$

where $\hat{\gamma}$ is a positive constant. With $u \in \mathbb{R}^{n_-}$, $z \in \mathbb{R}^{n-n_-}$ we may write system (A.22) in decoupled form

$$\dot{u} = A^{-}(t)u \tag{A.23}$$

$$\dot{z} = A'(t)z \tag{A.24}$$

and for the fundamental matrices $\Psi^{-}(t)$, $\Psi'(t)$ of (A.23), (A.24), respectively, we get

$$\| \Psi^{-}(t)(\Psi^{-})^{-1}(s) \| \leq 2^{1/2} K \hat{\gamma} e^{-\alpha(t-s)} \quad \text{for all } t \geq s \geq 0,$$

$$\| \Psi'(t)D'(\Psi')^{-1}(s) \| \leq 2^{1/2} K \hat{\gamma} e^{-\beta(s-t)} \quad \text{for all } s \geq t \geq 0, \tag{A.25}$$

where $D' := \mathrm{diag}(I_{n_+}, O_{n - n_- - n_+})$. Now we can apply the same reduction procedure as above to system (A.24), keeping in mind the estimate (A.25) for the corresponding fundamental matrix $\Psi'(t)$. Thus, by means of a second transformation (possible by Lemmas A.1 and A.2) of the form

$$\begin{pmatrix} v \\ w \end{pmatrix} = S_2(t)z, \quad v \in \mathbb{R}^{n_+}, \quad w \in \mathbb{R}^{n-n_--n_+},$$

we reduce system (A.24) to a decoupled system

$$\dot{v} = A^{+}(t)v \tag{A.26}$$

$$\dot{w} = A^{*}(t)w.$$

Here we get for the fundamental matrix $\Psi^{+}(t)$ (= left upper block of $\Psi'(t)$) of (A.26) the estimate

$$\| \Psi^{+}(t)(\Psi^{+})^{-1}(s) \| \leq 2K K' \hat{\gamma} e^{-\beta(s-t)} \quad \text{for all } s \geq t \geq 0,$$

where K' is an upper bound for $\| S_2^{-1}(t) \|$ on $[0,\infty)$. Hence, Theorem A.1 is proved. The transformation matrix $T(t)$ is explicitly given by $\mathrm{diag}(I_{n_-}, S_2(t))\, S_1(t)$. ∎

Appendix B: Linear Equations

It is the aim of this appendix to prove a couple of lemmas on linear equations of a particular form (Lemmas B.2 and B.6) which play an essential role in the derivation of the main results of this book.

B.I: Differential Equations

In the continuous time case the proof of our Lemma B.2 below rests on a well known perturbation result which we take from the literature. Its proof is based on a straightforward application of Gronwall's inequality and is therefore omitted (see e.g. Coppel [12, Proposition 1.1]). Consider a linear differential system

$$\dot{x} = A(t)x \tag{B.1}$$

and a linear perturbation

$$\dot{x} = [A(t) + B(t)]x. \tag{B.2}$$

We denote the respective principal fundamental matrices by $\Phi(t,s)$ and $\Psi(t,s)$.

Lemma B.1: Suppose that $A(t)$ and $B(t)$ in (B.2) are real square matrices whose elements are continuous on an interval J. Furthermore, suppose that there exist positive numbers α,γ,δ such that $\| B(t) \| \leq \delta$

on the interval J and

$$\|\Phi(t,s)\| \leq \gamma e^{-\alpha(t-s)} \quad \text{for all } t,s \in J,\ t \geq s.$$

Then

$$\|\Psi(t,s)\| \leq \gamma e^{(-\alpha+\delta\gamma)(t-s)} \quad \text{for all } t,s \in J,\ t \geq s.$$

In the next lemma we consider a particular class of perturbations of the decoupled linear system

$$\dot{u} = A^-(t)u$$
$$\dot{v} = A^+(t)v$$
$$\dot{w} = 0$$

where the matrices $A^-(t)$, $A^+(t)$ are supposed to be continuous and bounded for all $t \geq 0$. Let α,γ be positive constants such that the principal fundamental matrices $\Phi^-(t,s)$, $\Phi^+(t,s)$ of $\dot{u} = A^-(t)u$, $\dot{v} = A^+(t)v$, respectively, satisfy the estimates

$$\|\Phi^-(t,s)\| \leq \gamma e^{-\alpha(t-s)} \quad \text{for all } t \geq s \geq 0,$$

$$\|\Phi^+(t,s)\| \leq \gamma e^{-\alpha(s-t)} \quad \text{for all } s \geq t \geq 0.$$

<u>Lemma B.2</u>: For fixed matrices $A^-(t)$, $A^+(t)$ as above consider the class of linear differential systems

$$\begin{aligned} \dot{u} &= [A^-(t) + B_1(t)]u + B_2(t)v \\ \dot{v} &= [A^+(t) + B_3(t)]v \\ \dot{w} &= B_4(t)u + B_5(t)v \end{aligned} \qquad \text{(B.3)}$$

where the matrices $B_i(t)$, $i = 1,\dots,5$ are continuous and bounded (in norm) above by $\frac{\alpha}{2\gamma}$ on a compact interval $J_o := [t_o, T_o]$, $0 \leq t_o \leq T_o$.

Then any solution $(u(t), v(t), w(t))$ of system (B.3) satisfies the inequality

$$\|w(T_o)\| \leq \|w(t_o)\| + \|u(t_o)\| + \frac{3}{2}\|v(T_o)\| .$$

Remark: The crucial point of this lemma is that the asserted inequality holds true uniformly for any solution of any system of the form (B.3) as long as the matrices $B_i(t)$ are bounded by $\frac{\alpha}{2\gamma}$. This bound depends only on the matrices $A^-(t)$ and $A^+(t)$, in particular it is independent of the length of the interval J_o.

Proof of Lemma B.2: Later in this proof we need the following estimates which are easily verified:

$$\int_{t_o}^{T_o} e^{-\frac{\alpha}{2}(\tau - t_o)} d\tau \leq \frac{2}{\alpha}, \qquad \int_{t_o}^{T_o} e^{\frac{\alpha}{2}(\tau - T_o)} d\tau \leq \frac{2}{\alpha},$$

$$\int_{t_o}^{T_o} \int_{t_o}^{\tau} e^{-\frac{\alpha}{2}(\tau-\sigma)} e^{\frac{\alpha}{2}(\sigma - T_o)} d\sigma d\tau \leq \frac{2}{\alpha^2} .$$

By Lemma B.1 we get for the principal fundamental matrices $\Psi^-(t,s)$ and $\Psi^+(t,s)$ of $\dot{u} = [A^-(t) + B_1(t)]u$ and $\dot{v} = [A^+(t) + B_3(t)]v$, respectively, the estimates

$$\|\Psi^-(t,s)\| \leq \gamma e^{-\frac{\alpha}{2}(t-s)} \quad \text{for all } t,s \in J_o,\ t \geq s,$$

$$\|\Psi^+(t,s)\| \leq \gamma e^{\frac{\alpha}{2}(t-s)} \quad \text{for all } t,s \in J_o,\ s \geq t.$$

For the components of any solution $(u(t), v(t), w(t))$ of system (B.3) we get for each $t \in J_o$ the relations

$$u(t) = \Psi^-(t,t_o)u(t_o) + \int_{t_o}^{t} \Psi^-(t,\sigma)B_2(\sigma)v(\sigma)\,d\sigma,$$

$$v(t) = \Psi^+(t,T_o)v(T_o),$$

$$w(t) = w(t_o) + \int_{t_o}^{t} [B_4(\tau)u(\tau) + B_5(\tau)v(\tau)]\,d\tau,$$

whose combination leads to

$$w(T_o) = w(t_o) + \int_{t_o}^{T_o} B_4(\tau)\Psi^-(\tau,t_o)u(t_o)\,d\tau +$$

$$\int_{t_o}^{T_o}\int_{t_o}^{\tau} B_4(\tau)\Psi^-(\tau,\sigma)B_2(\sigma)\Psi^+(\sigma,T_o)v(T_o)\,d\sigma d\tau +$$

$$\int_{t_o}^{T_o} B_5(\tau)\Psi^+(\tau,T_o)v(T_o)\,d\tau.$$

This in turn yields the estimate

$$\| w(T_o)\| \le \| w(t_o)\| + \frac{\alpha}{2}\| u(t_o)\| \int_{t_o}^{T_o} e^{-\frac{\alpha}{2}(\tau-t_o)}\,d\tau\ -$$

$$\frac{\alpha^2}{4}\| v(T_o)\| \int_{t_o}^{T_o}\int_{t_o}^{\tau} e^{-\frac{\alpha}{2}(\tau-\sigma)}\ e^{\frac{\alpha}{2}(\sigma-T_o)}\,d\sigma d\tau +$$

$$\frac{\alpha}{2}\| v(T_o)\| \int_{t_o}^{T_o} e^{\frac{\alpha}{2}(\tau-T_o)}\,d\tau.$$

Application of the above inequalities proves the lemma. ∎

B.II: Difference Equations

The treatment of the discrete time case goes parallel to the continuous time case. However, since the qualitative theory of linear difference equations is somewhat underdeveloped compared to the theory of differential equations we cannot quote the discrete analogue (Lemma B.5 with Corollary) of Lemma B.1 from the literature. In order to prove this auxiliary result which is of interest on its own right we therefore state such basic concepts as the variation of constants formula and Gronwall's inequality for difference equations. Due to the time-asymmetry inherent in difference equations each of these results appears in two versions, one for each time direction.

Consider the linear inhomogeneous difference equation

$$x(k+1) = A(k)x(k) + b(k) \qquad (B.4)$$

where the vectors $x(k)$, $b(k)$ and the matrix $A(k)$ are defined for all integers k. Let $\Phi(k,l)$, k,l integer, denote the principal fundamental matrix of the homogeneous equation associated with (B.4), meaning that

$$\Phi(k+1,l) = A(k)\Phi(k,l), \quad \Phi(l,l) = I, \text{ for all } k \geq l+1.$$

(Notice the typographical difference between the letter "l" and the numeral "1".)

Our first lemma concerns the variation of constants formula. Its proof is straightforward and therefore omitted.

<u>Lemma B.3</u>: (i) The unique (forward) solution of equation (B.4) with initial value $x(l)$ can be represented for $k \geq l+1$ in the form

$$x(k) = \Phi(k,l)x(l) + \sum_{i=l}^{k-1} \Phi(k,i+1)b(i).$$

(ii) If $A(k)$ is nonsingular for each integer k then the unique (backward) solution of equation (B.4) with initial value $x(l)$ can be represented for $k \leq l-1$ in the form

$$x(k) = \Phi(k,l)x(l) - \sum_{i=k+1}^{l} \Phi(k,i)b(i-1).$$

The following lemma will later be used in the same sense as Gronwall's Lemma is used in the theory of differential equations.

<u>Lemma B.4</u>: Let c and d be arbitrary real numbers and let l be any integer. Then the following holds true:

(i) If a real sequence a_k, $k \geq l$, satisfies the implicit relation

$$a_k \leq c + d \sum_{i=l}^{k-1} a_i \quad \text{for all } k \geq l+1, \tag{B.5}$$

then it satisfies the explicit estimate

$$a_k \leq (c + da_l)(1+d)^{k-l-1} \quad \text{for all } k \geq l+1. \tag{B.6}$$

(ii) Suppose in addition that $d < 1$. Then, if a real sequence a_k, $k \leq l$, satisfies the implicit relation

$$a_k \leq c + d \sum_{i=k+1}^{l} a_{i-1} \quad \text{for all } k \leq l-1, \tag{B.7}$$

it satisfies the explicit estimate

$$a_k \leq c(1-d)^{k-1} \text{ for all } k \leq l-1. \tag{B.8}$$

Proof: (i) For $k = l+1$ the relations (B.5) and (B.6) are identical. Assuming the validity of the estimate $a_i \leq (c+da_l)(1+d)^{i-l-1}$ for all i with $l+1 \leq i \leq k$ we conclude that

$$a_{k+1} \leq c + da_l + d \sum_{i=l+1}^{k} a_i \leq (c+da_l) + d \sum_{i=l+1}^{k} (c+da_l)(1+d)^{i-l-1} =$$

$$(c+da_l)\,[\,1 + d \sum_{j=0}^{k-l-1} (1+d)^j\,]\,.$$

A straightforward induction argument shows that the last expression equals $(c+da_l)(1+d)^{k-l}$.

(ii) The first two steps of the induction are easily verified. Assuming the validity of the estimate $a_i \leq c(1-d)^{i-l}$ for all i between $k+1$ and $l-1$ we get

$$a_k \leq c + da_k + d \sum_{i=k+1}^{l-1} a_i \leq c + da_k + d \sum_{i=k+1}^{l-1} c(1-d)^{i-l}$$

and from this

$$a_k(1-d) \leq c\,[\,1 + d \sum_{i=1}^{l-k-1} (1-d)^{-i}\,]\,.$$

A straightforward induction completes the proof. ■

Next we prove a perturbation result which is the discrete analogue of Lemma B.1. Consider the homogeneous difference equation

$$x(k+1) = A(k)x(k) \tag{B.9}$$

and a perturbation

$$x(k+1) = [A(k) + B(k)]x(k) \tag{B.10}$$

whose respective principal fundamental matrices are denoted by $\Phi(k,l)$ and $\Psi(k,l)$.

<u>Lemma B.5</u>: Suppose that equation (B.10) is defined for all k from a set J of consecutive integers. Furthermore suppose that $\| B(k) \|$ is bounded above on J by some positive constant δ. Then the following is true:

(i) If there exist positive constants γ, λ such that

$$\| \Phi(k,l) \| \leq \gamma \lambda^{k-l} \quad \text{for all } k,l \in J,\ k \geq l+1$$

then

$$\| \Psi(k,l) \| \leq \beta \mu^{k-l} \quad \text{for all } k,l \in J,\ k \geq l+1$$

with $\beta := (\gamma\lambda + \gamma\delta)/(\lambda + \gamma\delta)$ and $\mu := \lambda + \gamma\delta$.

(ii) If A(k) is nonsingular for each $k \in J$ and if there exist positive constants $\tilde{\gamma}, \tilde{\lambda}$ with $(\tilde{\gamma}\tilde{\lambda})^{-1} > \delta$ such that

$$\| \Phi(k,l) \| \leq \tilde{\gamma} \tilde{\lambda}^{l-k} \quad \text{for all } k,l \in J,\ k \leq l-1$$

then

$$\| \Psi(k,l) \| \leq \tilde{\beta} \tilde{\mu}^{l-k} \quad \text{for all } k,l \in J,\ k \leq l-1$$

with $\tilde{\beta} := \tilde{\gamma}$ and $\tilde{\mu} := \tilde{\lambda}/(1 - \tilde{\gamma}\delta\tilde{\lambda})$.

Corollary: (i) If in case (i) of Lemma B.5 the constant λ is strictly between 0 and 1, i.e. (B.9) is asymptotically stable, then with the choice $\delta < (1-\lambda)/\gamma$ also μ is strictly between 0 and 1, i.e. also (B.10) is asymptotically stable. In particular,

$$\mu = \frac{1+\lambda}{2} \quad \text{if} \quad \delta = \frac{1-\lambda}{2\gamma}.$$

(ii) If in case (ii) of Lemma B.5 the constant $\tilde{\lambda}$ is strictly between 0 and 1, i.e. (B.9) is asymptotically stable in negative time, then with the choice $\delta < (1-\tilde{\lambda})/\tilde{\gamma}\tilde{\lambda}$ also $\tilde{\mu}$ is strictly between 0 and 1, i.e. also (B.10) is asymptotically stable in negative time. In particular,

$$\tilde{\mu} = \frac{1+\tilde{\lambda}}{2} \quad \text{if} \quad \delta = \frac{1-\tilde{\lambda}}{\tilde{\lambda}\tilde{\gamma}(1+\tilde{\lambda})}.$$

Proof of Lemma B.5: (i) By Lemma B.3 an arbitrary forward solution of equation (B.10) can be written in the form

$$x(k) = \Phi(k,l)x(l) + \sum_{i=l}^{k-1} \Phi(k,i+1)B(i)x(i) \quad \text{for } k \geq l+1$$

which implies the estimate

$$\|x(k)\| \leq \gamma\lambda^{k-l}\|x(l)\| + \gamma\delta \sum_{i=l}^{k-1} \lambda^{k-i-1}\|x(i)\| \quad \text{for all } k,l \in J,\ k \geq l+1.$$

Next we apply Lemma B.4 to the relation

$$\lambda^{-k}\|x(k)\| \leq \gamma\lambda^{-l}\|x(l)\| + \gamma\delta\lambda^{-1} \sum_{i=l}^{k-1} \lambda^{-i}\|x(i)\|$$

yielding

$$\lambda^{-k}\|x(k)\| \leq \lambda^{-l}\|x(l)\|(\gamma + \gamma\delta\lambda^{-1})(1+\gamma\delta\lambda^{-1})^{k-l-1},$$

in rewritten form

$$\| x(k) \| \leq \frac{\gamma\lambda + \gamma\delta}{\lambda + \gamma\delta} \| x(1) \| (\lambda + \gamma\delta)^{k-1} \quad \text{for all } k,1 \in J,\ k \geq 1+1.$$

Because of the relation $\| \Psi(k,1) \| = \sup\{ \| \Psi(k,1)\xi \| / \| \xi \| : \xi \neq 0\}$ we get from the last inequality the claimed estimate.

(ii) By Lemma B.3 an arbitrary backward solution of equation (B.10) can be written in the form

$$x(k) = \Phi(k,1)x(1) - \sum_{i=k+1}^{1} \Phi(k,i)B(i-1)x(i-1) \quad \text{for all } k \leq 1-1$$

implying

$$\tilde{\lambda}^{k} \| x(k) \| \leq \tilde{\gamma}\tilde{\lambda}^{1} \| x(1) \| + \tilde{\gamma}\delta\tilde{\lambda} \sum_{i=k+1}^{1} \tilde{\lambda}^{i-1} \|x(i-1)\| \quad \text{for all } k,1 \in J,\ k \leq 1$$

Using Lemma B.4 (note that $\tilde{\gamma}\delta\tilde{\lambda} < 1$) we get

$$\tilde{\lambda}^{k} \| x(k) \| \leq \tilde{\gamma}\tilde{\lambda}^{1} \| x(1) \| (1 - \tilde{\gamma}\delta\tilde{\lambda})^{k-1} \quad \text{for all } k,1 \in J,\ k \leq 1-1$$

and from this

$$\| \Psi(k,1) \| \leq \tilde{\gamma} \left(\frac{\tilde{\lambda}}{1 - \tilde{\gamma}\delta\tilde{\lambda}}\right)^{1-k} \quad \text{for all } k,1 \in J,\ k \leq 1-1.$$

Thus, the proof of Lemma B.5 is complete. ■

The proof of the Corollary to Lemma B.5 requires only a couple of elementary calculations which are omitted.

Now we are in a position to prove the main auxiliary result for the discrete time case. As unperturbed equation we consider the decoupled linear system

$$\begin{aligned} u(k+1) &= A^-(k)u(k) \\ v(k+1) &= A^+(k)v(k) \\ w(k+1) &= w(k), \end{aligned}$$

where the matrices $A^-(k)$, $A^+(k)$ are defined (and $A^+(k)$ is nonsingular) for all nonnegative integers k. We suppose that the principal fundamental matrices $\Phi^-(k,l)$, $\Phi^+(k,l)$ of $u(k+1) = A^-(k)u(k)$ and $v(k+1) = A^+(k)v(k)$, respectively, satisfy

$$\|\Phi^-(k,l)\| \leq \gamma\,\lambda^{k-l} \quad \text{for all } k \geq l \geq 0,$$

$$\|\Phi^+(k,l)\| \leq \gamma\,\lambda^{l-k} \quad \text{for all } l \geq k \geq 0,$$

where the constants γ,λ obey the relations

$$\gamma \geq 1, \quad 0 < \lambda < 1.$$

Lemma B.6: For fixed matrices $A^-(k)$, $A^+(k)$ as above consider any system of difference equations of the form

$$\begin{aligned} u(k+1) &= [A^-(k) + B_1(k)]u(k) + B_2(k)v(k) \\ v(k+1) &= [A^+(k) + B_3(k)]v(k) \\ w(k+1) &= w(k) + B_4(k)u(k) + B_5(k)v(k) \end{aligned} \qquad (B.11)$$

where the matrices $B_i(k)$, $i = 1,\dots,5$ are defined and bounded (in norm) above by $\frac{(1-\lambda)}{2\gamma}$ on a set

$$J_o := \{k_o, k_o+1, \dots, K_o-1\}$$

of consecutive integers.

Then any solution $(u(k),v(k),w(k))$ of (B.11) on J_o satisfies the estimate

$$\|w(K_o)\| \leq \|w(k_o)\| + \|u(k_o)\| + 2\|v(K_o)\|. \tag{B.12}$$

Remark: As in the continuous time case (Lemma B.2) the main point of this lemma is that the estimate (B.12) is valid for the whole class of systems of the form (B.11) as long as the boundedness conditions for the matrices $B_i(k)$ are fulfilled. In particular the "length" of J_o has no influence on the coefficients occuring in the inequality (B.12).

Proof: The upper bound $\frac{(1-\lambda)}{2\gamma}$ for the matrices $B_1(k)$ and $B_3(k)$ allows us to apply either part of the Corollary to Lemma B.5 with the particular choice of δ stated there (note that $(1-\lambda)/2\gamma < (1-\lambda)/(\lambda\gamma + \lambda^2\gamma)$). This yields for the principal fundamental matrices $\Psi^-(k,l)$, $\Psi^+(k,l)$ of $u(k+1) = [A^-(k) + B_1(k)]u(k)$, $v(k+1) = [A^+(k) + B_3(k)]v(k)$, respectively, the estimates

$$\|\Psi^-(k,l)\| \leq \gamma\left(\frac{1+\lambda}{2}\right)^{k-l} \quad \text{for all } k,l \in J_o,\ k \geq l+1, \tag{B.13}$$

$$\|\Psi^+(k,l)\| \leq \gamma\left(\frac{1+\lambda}{2}\right)^{l-k} \quad \text{for all } k,l \in J_o,\ k \leq l-1 \tag{B.14}$$

(note that $(2\gamma\lambda + 1 - \lambda)/(1+\lambda) \leq \gamma$). Now let $(u(k),v(k),w(k))$ be any solution of equation (B.11) on J_o. Then we get

$$u(s) = \Psi^-(s,r)u(r) + \sum_{i=r}^{s-1} \Psi^-(s,i+1)B_2(i)v(i) \quad \text{for } s \geq r+1,$$

$$v(s) = \Psi^+(s,r)v(r) \quad \text{for all } r \text{ and } s. \tag{B.15}$$

If $K_o = k_o + 1$, i.e. J_o is a singleton, we get from (B.11) immediately

$$\|w(k_o+1)\| \leq \|w(k_o)\| + \frac{1-\lambda}{2\gamma}\|u(k_o)\| + \frac{1-\lambda}{2\gamma}\|v(k_o)\|$$

and with (B.15) and (B.14)

$$\|w(K_o)\| \leq \|w(k_o)\| + \frac{1-\lambda}{2\gamma}\|u(k_o)\| + \frac{1-\lambda^2}{4}\|v(K_o)\| .$$

This obviously implies the asserted estimate in the particular case $K_o = k_o + 1$. In the general case $K_o \geq k_o + 2$ we get with Lemma B.3

$$w(K_o) = w(k_o) + \sum_{j=k_o}^{K_o-1} [B_4(j)u(j) + B_5(j)v(j)] =$$

$$w(k_o) + [\sum_{j=k_o}^{K_o-1} B_5(j)\Psi^+(j,K_o)]v(K_o) + B_4(k_o)u(k_o) +$$

$$\sum_{j=k_o+1}^{K_o-1} B_4(j)\Psi^-(j,k_o)u(k_o) +$$

$$\sum_{j=k_o+1}^{K_o-1} B_4(j) \sum_{i=k_o}^{j-1} \Psi^-(j,i+1)B_2(i)\Psi^+(i,K_o)v(K_o)$$

and from this with (B.13) and (B.14) (using the geometric series with $(1+\lambda)/2 \in (0,1)$)

$$\|w(K_o)\| \leq \|w(k_o)\| + \frac{1-\lambda}{2}\|v(K_o)\| \sum_{j=k_o}^{K_o-1} (\frac{1+\lambda}{2})^{K_o-j} + \frac{1-\lambda}{2\gamma}\|u(k_o)\| +$$

$$\frac{1-\lambda}{2}\|u(k_o)\| \sum_{j=k_o+1}^{K_o-1} (\frac{1+\lambda}{2})^{j-k_o} +$$

$$(\frac{1-\lambda}{2})^2 \|v(K_o)\| \sum_{j=k_o+1}^{K_o-1} \sum_{i=k_o}^{j-1} (\frac{1+\lambda}{2})^{K_o+j-2i-1} \leq$$

$$\|w(k_o)\| + \frac{1-\lambda}{2}(1-\frac{1+\lambda}{2})^{-1}\|v(K_o)\| +$$

$$(\frac{1-\lambda}{2\gamma} + \frac{1-\lambda^2}{4}(1 - \frac{1+\lambda}{2})^{-1}) \|u(k_o)\| \ +$$

$$(\frac{1-\lambda}{2})^2 \|v(K_o)\| \ (\frac{1+\lambda}{2})^{K_o-1} \sum_{j=k_o+1}^{K_o-1} (\frac{1+\lambda}{2})^j \sum_{i=k_o}^{j-1} (\frac{1+\lambda}{2})^{-2i} .$$

The sum of the first three terms in the last expression has upper bound

$$\|w(k_o)\| + \|v(K_o)\| + \|u(k_o)\|$$

whereas the essential part of the double sum can be estimated in the following way (temporarily abbreviating $(1+\lambda)/2$ by ε):

$$\varepsilon^{K_o-1} \sum_{j=k_o+1}^{K_o-1} \varepsilon^j \frac{\varepsilon^{-2k_o} - \varepsilon^{-2j}}{1-\varepsilon^{-2}} =$$

$$\frac{\varepsilon^{K_o-1}}{1-\varepsilon^{-2}} [\varepsilon^{-2k_o} \frac{\varepsilon^{k_o+1} - \varepsilon^{K_o}}{1-\varepsilon} - \frac{\varepsilon^{-k_o-1} - \varepsilon^{-K_o}}{1-\varepsilon^{-1}}] =$$

$$\frac{1}{(1-\varepsilon^{-2})(1-\varepsilon)} [\varepsilon^{K_o-k_o} - \varepsilon^{2K_o-2k_o-1} + \varepsilon^{K_o-k_o-1} - 1] =$$

$$\frac{\varepsilon^2}{(1-\varepsilon^2)(1-\varepsilon)} (1 - \varepsilon^{K_o-k_o-1})(1 - \varepsilon^{K_o-k_o}) <$$

$$\frac{\varepsilon^2}{(1-\varepsilon^2)(1-\varepsilon)} < (\frac{\varepsilon}{1-\varepsilon})^2 = (\frac{1+\lambda}{1-\lambda})^2 .$$

With this the asserted inequality (B.12) follows readily. ∎

Appendix C: An Algebraic Equation arising in Population Genetics

From Sections 6 and 11 it became apparent that the continuous time and the discrete time equation of the Fisher-Wright-Haldane model bear much similarity. This resemblance is particularly striking when one compares the sets of stationary solutions for both equations. Essentially, the stationary solutions in either case turn out to be the solutions of one particular algebraic equation ((C.1) below) which is the object to be investigated in this appendix. The precise knowledge of those solution sets is required for the analysis of the convergence problem (see Sections 6 and 11).

Let $F = (f_{rs})$ be any symmetric $n\times n$ matrix with nonnegative elements and let

$$\Phi(x) \; := x^T F x, \; x \in \mathbb{R}^n ,$$

be the corresponding quadratic form. In this appendix we study the system of algebraic equations

$$\begin{pmatrix} x_1 & & 0 \\ & \ddots & \\ 0 & & x_n \end{pmatrix} F \begin{pmatrix} x_1 \\ \vdots \\ x_n \end{pmatrix} = \Phi(x) \begin{pmatrix} x_1 \\ \vdots \\ x_n \end{pmatrix} . \tag{C.1}$$

It is our goal to describe the solution set

$$E \; := \{x \in \mathbb{R}^n_+ : \; x \text{ satisfies } (C.1)\}$$

of system (C.1) in the nonnegative orthant

$$\mathbb{R}^n_+ := \{x \in \mathbb{R}^n : x_i \geq 0, \ i = 1, \ldots, n\}$$

of $\mathbb{R}^n$. It will turn out that the structure of the solution set E depends in a very particular way on the structure of the matrix F. In order to take care of this fact it is advisable to decompose equation (C.1) in a suitable way. To this end we recall the notation already used before. Let

$$N := \{1, \ldots, n\}$$

be the set of the first n positive integers and denote by $N_1, \ldots, N_{2^n}$ (for the moment in any order) the different elements of the power set of N, for definiteness set $N_{2^n} := \emptyset$. For $j = 1, \ldots, 2^n$ define

$$n_j := \text{card } N_j,$$

the cardinal number of N_j.

We first decompose the matrix F. To each $j \in \{1, \ldots, 2^n - 1\}$ we associate the $n_j \times n_j$ submatrix of F

$$F_j := (f_{r_\nu s_\nu}), \quad \nu = 1, \ldots, n_j,$$

where the r_ν, s_ν are elements of N_j such that $r_1 < \ldots\ldots < r_{n_j}$ and $s_1 < \ldots\ldots < s_{n_j}$. Corresponding to each F_j we define the number

$$d_j := \dim \ker F_j,$$

the dimension of the kernel of F_j, $d_j \leq n_j$. Without loss of generality we may now index the subsets $N_1, \ldots, N_{2^n-1}$ in such a way that

$$\begin{aligned} d_j &< n_j \quad \text{for } j = 1,\dots,n_o, \\ d_j &= n_j \quad \text{for } j = n_o+1,\dots,2^n-1, \end{aligned} \tag{C.2}$$

where n_o is some number between 1 and 2^n-1.

Next we decompose the nonnegative orthant $\mathbb{R}^n_+$ of $\mathbb{R}^n$. $\mathbb{R}^n_+$ may be represented as the disjoint union

$$\mathbb{R}^n_+ = \bigcup_{j=1}^{2^n} R_j \tag{C.3}$$

where

$$R_j := \{x \in \mathbb{R}^n_+ : x_i > 0 \text{ for } i \in N_j,\ x_k = 0 \text{ for } k \in N \setminus N_j\}.$$

For $j \neq 2^n$, R_j is the positive orthant of the n_j-dimensional subspace

$$S_j := \{x \in \mathbb{R}^n : x_k = 0 \text{ for } k \in N \setminus N_j\}$$

of $\mathbb{R}^n$. For later purposes it is useful to introduce the canonical isomorphism

$$J_j : S_j \to \mathbb{R}^{n_j},$$

the effect of which is just to drop the zero components of the vectors in S_j. Denoting the points in $\mathbb{R}^{n_j}$ by x^j we define on $\mathbb{R}^{n_j}$ the quadratic form

$$\Phi_j(x^j) := (x^j)^T F_j x^j$$

and notice its relation

$$\Phi_j(J_j(x)) = \Phi(x) \text{ for all } x \in \mathbb{R}^n$$

to the quadratic form Φ on $\mathbb{R}^n$.

Now we are prepared to investigate the solution set E of equation (C.1) in $\mathbb{R}^n_+$. We do this on each subset R_j of $\mathbb{R}^n_+$ separately and so we now fix any $j \in \{1,\dots,2^n-1\}$. For $x \in R_j$ those of the equations in system (C.1) which have indices in $N \setminus N_j$ are trivially satisfied (both sides vanish) and hence, via the relation

$$x^j := J_j(x),$$

equation (C.1) is equivalent (on R_j) to the equation

$$\operatorname{diag}(x^j)F_j x^j = \Phi_j(x^j)x^j. \tag{C.4}$$

The relation between the solution set

$$E_j := \{x^j \in \overset{\circ}{\mathbb{R}}{}^{n_j}_+ : x^j \text{ satisfies (C.4)}\}$$

of (C.4) in the interior $\overset{\circ}{\mathbb{R}}{}^{n_j}_+$ of $\mathbb{R}^{n_j}_+$ and $E \cap R_j$ can be expressed by means of the isomorphism J_j as

$$J_j(E \cap R_j) = E_j.$$

Since $x^j_i \neq 0$ for $i = 1,\dots,n_j$ equation (C.4), on the other hand, is equivalent to

$$F_j x^j = \Phi_j(x^j)e^j \tag{C.5}$$

where $e^j := (1,\dots,1)^T \in \mathbb{R}^{n_j}$. Hence, E_j may be characterized as

$$E_j = \{x^j \in \mathbb{R}^{n_j}_+ : F_j x^j = \Phi_j(x^j)e^j,\ x^j_i > 0 \text{ for } i = 1,\dots,n_j\}.$$

Now we distinguish two cases.

Case 1: $j \geq n_o+1$ (see (C.2)). In this case F_j is the $n_j \times n_j$ zero matrix. This means that $\Phi_j(x^j) = 0$ for all $x^j \in \overset{o}{\mathbb{R}}^{n_j}$ and that $E_j = \overset{o}{\mathbb{R}}{}_+^{n_j}$. So we get the following lemma on this trivial case.

Lemma C.1: If $j \geq n_o+1$, i.e. $d_j = n_j$, then

(i) $$F_j = O_{n_j \times n_j},$$

(ii) $$E_j = \overset{o}{\mathbb{R}}{}_+^{n_j},$$

(iii) $$\Phi_j(x^j) \equiv 0 \text{ on } \overset{o}{\mathbb{R}}{}_+^{n_j}.$$

Case 2: $j \leq n_o$. In this case F_j is different from the zero matrix. On the other hand, all elements of F are nonnegative. Thus, for each x^j in $\overset{o}{\mathbb{R}}{}_+^{n_j}$ the vector $F_j x^j$ has nonnegative components, not all of them vanishing. This implies that $(x^j)^T F_j x^j$ is positive, i.e.

$$\Phi_j(x^j) > 0 \text{ on } \overset{o}{\mathbb{R}}{}_+^{n_j} \text{ for all } j \leq n_o.$$

That in this case Φ_j does not vanish in $\overset{o}{\mathbb{R}}{}_+^{n_j}$ is crucial for the following trick to work. With the nonlinear equation (C.5) that we want to solve we associate the linear system

$$F_j x^j = e^j$$

whose solution set L_j in $\overset{o}{\mathbb{R}}{}_+^{n_j}$,

$$L_j := \{x^j \in \overset{o}{\mathbb{R}}{}_+^{n_j} : F_j x^j = e^j\},$$

is the (possibly empty) intersection of an affine space of dimension d_j

with $\overset{\circ}{\mathbb{R}}_{+}^{n_j}$. In order to relate this smooth manifold L_j to the solution set E_j of (C.5) we introduce the smooth map D_j from $\overset{\circ}{\mathbb{R}}_{+}^{n_j}$ into itself via

$$D_j(x^j) := \Phi_j(x^j)^{-1}x^j.$$

From this definition it is immediate that

$$\Phi_j(D_j(x^j)) = \Phi_j(x^j)^{-1} \text{ for all } x^j \in \overset{\circ}{\mathbb{R}}_{+}^{n_j}. \tag{C.6}$$

The role of the map D_j is described in the next lemma.

<u>Lemma C.2</u>: D_j maps L_j diffeomorphically onto E_j.

<u>Proof</u>: a) For $x^j \in L_j$ we get with (C.6)

$$F_jD_j(x^j) = \Phi_j(x^j)^{-1}F_jx^j = \Phi_j(x^j)^{-1}e^j = \Phi_j(D_j(x^j))e^j$$

and this means that $D_j(x^j) \in E_j$.

b) For $x^j \in E_j$ we get

$$F_jD_j(x^j) = \Phi_j(x^j)^{-1}F_jx^j = e^j$$

and this means that $D_j(x^j) \in L_j$.

c) From the definition of D_j and (C.6) it follows that

$$D_j^2(x^j) = D_j(D_j(x^j)) = \Phi_j(D_j(x^j))^{-1}D_j(x^j) = x^j$$

for all $x^j \in \overset{\circ}{\mathbb{R}}_{+}^{n_j}$. Thus, D_j^2 is the identity on $\overset{\circ}{\mathbb{R}}_{+}^{n_j}$ and with a) and b) we get

$$D_j(L_j) = E_j \text{ and } D_j(E_j) = L_j.$$

This completes the proof of Lemma C.2. ∎

We summarize what has been gained so far in case 2.

> <u>Lemma C.3</u>: If $j \leq n_o$, i.e. $d_j = \dim \ker F_j < n_j$, then
>
> (i) the solution set E_j of (C.4) in $\overset{o}{\mathbb{R}}{}^{n_j}_{+}$ is the (possibly empty) intersection of a d_j - dimensional affine subspace of $\mathbb{R}^{n_j}$ with $\overset{o}{\mathbb{R}}{}^{n_j}_{+}$, the positive orthant of $\mathbb{R}^{n_j}$,
>
> (ii) $E_j = D_j(L_j)$ is the (diffeomorphic) "projection" of the d_j - dimansional solution space L_j of $F_j x^j = e^j$ along rays in $\mathbb{R}^{n_j}_{+}$ through the coordinate origin into the hyperspace $\{x^j \in \overset{o}{\mathbb{R}}{}^{n_j} : \Sigma_{i=1}^{n_j} x_i^j = 1\}$,
>
> (iii) $\Phi_j(x^j) > 0$ on $\overset{o}{\mathbb{R}}{}^{n_j}_{+}$.

<u>Proof</u>: It remains to be shown that E_j is a linear manifold (see the footnote on the next page) and that the relation $\Sigma_{i=1}^{n_j} x_i^j = 1$ holds true for any $x^j \in E_j$. The latter relation is rather obvious since the inclusion $x_j \in E_j$ implies that

$$(x^j)^T F_j x^j = \Phi_j(x^j)(x^j)^T e^j$$

and with $\Phi_j(x^j) \neq 0$ we get from this the relation $(x^j)^T e^j = 1$. Next we show that E_j (if not empty) is the intersection of the affine space

$$x^j + \ker F_j$$

with $\overset{o}{\mathbb{R}}{}^{n_j}_{+}$ where x^j is some point of E_j. To this end it suffices to show that for any $y^j \in \ker F_j$ the vectors

$$x^j + \eta y^j, \quad \eta \in \mathbb{R},$$

are in E_j as long as they are positive. Hence, we have to prove the relation

$$F_j(x^j + \eta y^j) = (x^j + \eta y^j)^T F_j (x^j + \eta y^j) e^j$$

for all positive vectors $x^j + \eta y^j$. This, however, follows immediately from the relations $x^j \in E_j$, i.e. $F_j x^j = [(x^j)^T F_j x^j] e^j$, and $F_j y^j = 0$. ∎

Now we return to our problem of describing the solutions of equation (C.1) in $\mathbb{R}^n_+$.

<u>Theorem C.1</u>: (i) The solution set E of equation (C.1) in the nonnegative orthant $\mathbb{R}^n_+$ of $\mathbb{R}^n$ is the disjoint union

$$E = \bigcup_{j=1}^{2^n} (E \cap R_j)$$

where

a) for $j = 1, \ldots, n_o$ $E \cap R_j = J_j^{-1}(E_j) = J_j^{-1}(D_j(L_j))$ is empty or a linear manifold *) of dimension $d_j < n_j$ with $\Phi(x) > 0$ for all $x \in E \cap R_j$,

b) for $j = n_o+1, \ldots, 2^n-1$ the set $E \cap R_j = R_j$ is a linear manifold of dimension $d_j = n_j$ with $\Phi(x) = 0$ for all $x \in R_j$.

---

*) A manifold M is called *linear* if there exists an affine space A such that for any point on M there exists a neighborhood U such that $M \cap U = A \cap U$.

(ii) For $j = 1,\dots,n_o$ one has

$$E \cap R_j \subset \Lambda^{n-1} \quad (=\{x \in \mathbb{R}_+^n : \sum_{i=1}^{n} x_i = 1\}).$$

(iii) If $\Phi(x) = 0$ and $x \in \mathbb{R}_+^n$, then $x \in E$.

Proof: (i) and (ii) follow from the fact that the isomorphism J_j^{-1} "lifts" the properties of E_j in $\mathbb{R}_+^{n_j}$ to $E \cap R_j$ in $\mathbb{R}_+^n$. Next suppose that the assertion (iii) is not true. This means that for some $x_o \in \mathbb{R}_+^n$ we get the relations $\Phi(x_o) = 0$ and $x_o \notin E$ and herewith $\mathrm{diag}(x_o)Fx_o \neq 0$. With $x_o^j := J_j(x_o)$ where $j = j(x_o)$ is the index of x_o (see Section 6) we obtain $\mathrm{diag}(x_o^j)F_j x_o^j \neq 0$ and, because the components of x_o^j are positive, the vector $F_j x_o^j$ is nonnegative. Thus,

$$\Phi(x_o) = \Phi_j(x_o^j) = (x_o^j)^T F_j x_o^j$$

is a positive number. This is a contradiction. ∎

In our last theorem we give a sufficient condition for all solutions of (C.1) near a given solution to form a linear manifold.

Theorem C.2: Let $\bar{x}$ be a solution of (C.1) with

$$\bar{x} \in \Lambda^{n-1}.$$

Let $j = j(\bar{x})$ and $d = d(\bar{x})$ denote its index and defect, respectively (for the definitions see Section 6). Suppose that $N_j = \{1,\dots,n_j\}$ and that

$$\sum_{i=1}^{n_j} f_{ki}\bar{x}_i \neq \Phi(\bar{x}) \text{ for at least one subscript } k \in \{n_j+1,\ldots,n\}. \tag{C.7}$$

Then there exists a neighborhood U of $\bar{x}$ in $\mathbb{R}^n$ such that the set of solutions of equation (C.1) in $\Lambda^{n-1} \cap U$ forms a linear manifold of dimension d.

Proof: As before we partition

$$\bar{x} = \begin{pmatrix} \tilde{x} \\ 0 \end{pmatrix}, \quad F = \begin{pmatrix} F_j & F_{12} \\ F_{21} & F_{22} \end{pmatrix}$$

and notice that (C.7) in vector form reads

$$F_{21}\tilde{x} \neq \Phi(\bar{x})e^j \tag{C.8}$$

where $e^j = (1,\ldots,1)^T \in \mathbb{R}^{n_j}$. By Theorem C.1 there exists a d-dimensional linear manifold $M \subset R_j$ which contains all solutions of (C.1) in R_j. The claim of the theorem is that there exists a neighborhood U of $\bar{x}$ such that

$$E^{n-1} \cap U = M \cap U,$$

in other words, that $U \cap \Lambda^{n-1}$ contains no other solutions of (C.1) than those in M. Supposing the contrary means that there exists a sequence $\bar{x}_\nu$ of solutions of (C.1) in Λ^{n-1} but not in M, i.e. $\bar{x}_\nu \in E^{n-1} \setminus M$, such that

$$\lim_{\nu \to \infty} \bar{x}_\nu = \bar{x}. \tag{C.9}$$

Since the $\bar{x}_\nu$ are not in M, and hence not in R_j, each term of the sequence $\bar{x}_\nu$ must be in (exactly) one of the finitely many subsets $R_{\hat{j}} \cap \Lambda^{n-1}$ of Λ^{n-1} where $\hat{j} \neq j$. Thus, there exists (at least) one set $R_{\hat{j}_o} \cap \Lambda^{n-1}$ with

$$R_{\hat{j}_o} \cap R_j = \emptyset , \tag{C.10}$$

which contains an (infinite) subsequence of $\bar{x}_\nu$, again denoted by $\bar{x}_\nu$. This means that any component sequence of the vector sequence $\bar{x}_\nu$ is either a sequence of zeros or a sequence of positive numbers. Since $\bar{x}$ has $n_j \geq 1$ positive components it follows from (C.9) that n_j of the component sequences of $\bar{x}_\nu$ are eventually positive. The remaining component sequences cannot all be eventually zero because this would mean that $\bar{x}_\nu \in R_j$ for almost all ν ($\nu \geq \nu_o$, say) and this is impossible because of (C.10). Thus, there exist m ($1 \leq m \leq n-n_j$) additional subsequences of the component sequences of $\bar{x}_\nu$ (again denoted by $\bar{x}_\nu$) with eventually positive terms. Without loss of generality we may rearrange the components of the state vector $x \in \Lambda^{n-1} \subset \mathbb{R}^n$ to get the following situation:

$$\bar{x}_\nu = \begin{pmatrix} \tilde{x}_\nu \\ y_\nu \\ z_\nu \end{pmatrix} \xrightarrow[\nu \to \infty]{} \begin{pmatrix} \tilde{x} \\ y \\ z \end{pmatrix} = \bar{x} \tag{C.11}$$

where

$$\tilde{x} > 0 \text{ and } \tilde{x}_\nu > 0 \text{ for all } \nu \geq \nu_o \quad (\tilde{x},\ \tilde{x}_\nu \in \mathbb{R}^{n_j}) , \tag{C.12}$$

$$y = 0 \text{ and } y_\nu > 0 \text{ for all } \nu \geq \nu_o \quad (y,\ y_\nu \in \mathbb{R}^{m}) , \tag{C.13}$$

$$z = 0 \text{ and } z_\nu = 0 \text{ for all } \nu \geq \nu_o \quad (z,\ z_\nu \in \mathbb{R}^{n-n_j-m}) . \tag{C.14}$$

By assumption, all $\bar{x}_\nu$ are solutions of (C.1) and this means that we get with the corresponding partition of F the following relation which is valid for all $\nu \geq \nu_o$:

$$\mathrm{diag}(\tilde{x}_\nu, y_\nu, 0) \begin{pmatrix} F_j & F_{12} & F_{13} \\ F_{21} & F_{22} & F_{23} \\ F_{31} & F_{32} & F_{33} \end{pmatrix} \begin{pmatrix} \tilde{x}_\nu \\ y_\nu \\ 0 \end{pmatrix} =$$

$$[(\tilde{x}_\nu^T, y_\nu^T, 0) \begin{pmatrix} F_j & F_{12} & F_{13} \\ F_{21} & F_{22} & F_{23} \\ F_{31} & F_{32} & F_{33} \end{pmatrix} \begin{pmatrix} \tilde{x}_\nu \\ y_\nu \\ 0 \end{pmatrix}] \begin{pmatrix} \tilde{x}_\nu \\ y_\nu \\ 0 \end{pmatrix} .$$

From this we get for all $\nu \geqq \nu_o$

$$\mathrm{diag}(\tilde{x}_\nu, y_\nu) \begin{pmatrix} F_j & F_{12} \\ F_{21} & F_{22} \end{pmatrix} \begin{pmatrix} \tilde{x}_\nu \\ y_\nu \end{pmatrix} = [(\tilde{x}_\nu^T, y_\nu^T) \begin{pmatrix} F_j & F_{12} \\ F_{21} & F_{22} \end{pmatrix} \begin{pmatrix} \tilde{x}_\nu \\ y_\nu \end{pmatrix}] \begin{pmatrix} \tilde{x}_\nu \\ y_\nu \end{pmatrix}$$

or equivalently (notice (C.12) and (C.13)),

$$\begin{pmatrix} F_j & F_{12} \\ F_{21} & F_{22} \end{pmatrix} \begin{pmatrix} \tilde{x}_\nu \\ y_\nu \end{pmatrix} = [(\tilde{x}_\nu^T, y_\nu^T) \begin{pmatrix} F_j & F_{12} \\ F_{21} & F_{22} \end{pmatrix} \begin{pmatrix} \tilde{x}_\nu \\ y_\nu \end{pmatrix}] \begin{pmatrix} 1 \\ \vdots \\ 1 \end{pmatrix} \in \mathbb{R}^{n-n_j} . \qquad \text{(C.16)}$$

Letting ν tend to infinity in (C.16) and using (C.11), (C.12), (C.13) we conclude that

$$F_{21}\tilde{x} = [\tilde{x}^T F_j \tilde{x}] \begin{pmatrix} 1 \\ \vdots \\ 1 \end{pmatrix} \in \mathbb{R}^m$$

and this is a contradiction to (C.8) when we recall that $m \geqq 1$. ∎

References

[1] B. Aulbach, Asymptotic Amplitude and Phase for Isochronic Families of Periodic Solutions, in "Analytical and Numerical Approaches to Asymptotic Problems in Analysis" (eds. O. Axelson, L. S. Frank, A. Van der Sluis), 265 - 271, North Holland, Amsterdam 1981.

[2] B. Aulbach, Behavior of Solutions near Manifolds of Periodic Solutions. J. Diff. Eqns. 39 (1981), 345 - 377.

[3] B. Aulbach, Invariant Manifolds with Asymptotic Phase. Nonl. Anal. TMA 6 (1982), 817 - 827.

[4] B. Aulbach, A Reduction Principle for Nonautonomous Differential Equations. Arch. Math. 39 (1982), 217 - 232.

[5] B. Aulbach, Approach to Hyperbolic Manifolds of Stationary Solutions, in "Equadiff 82" (eds. H. W. Knobloch, K. Schmitt), 56 - 66, Lecture Notes in Mathematics 1017, Springer, Berlin, Heidelberg, New York 1983.

[6] B. Aulbach, Iterates Approaching Hyperbolic Manifolds of Fixed Points. J. reine angew. Math., to appear.

[7] B. Aulbach and K. P. Hadeler, Convergence to Equlibrium in the Classical Model of Population Genetics. J. Math. Anal. Appl., to appear.

[8] J. Carr, Applications of Centre Manifold Theory. Springer, Berlin, Heidelberg, New York 1981.

[9] L. Cesari, Asymptotic Behavior and Stability Problems in Ordinary Differential Equations. Springer, Berlin, Heidelberg, New York 1959.

[10] S. N. Chow and J. K. Hale, Methods of Bifurcation Theory. Springer, Berlin, Heidelberg, New York 1982.

[11] E. A. Coddington and N. Levinson, Theory of Ordinary Differential Equations. McGraw - Hill, New York, Toronto, London 1955.

[12] W. A. Coppel, Dichotomies in Stability Theory. Lecture Notes in Mathematics 629, Springer, Berlin, Heidelberg, New York 1978.

[13] J. F. Crow and M. Kimura, An Introduction to Population Genetics Theory. Harper and Row, New York 1970.

[14] A. W. F. Edwards, Foundations of Mathematical Genetics. Cambridge University Press, Cambridge, London, New York, Melbourne 1977.

[15] W. Feller, A Geometric Analysis of Fitness in Triply Allelic Systems. Math. Biosci. 5 (1974), 19 - 38.

[16] N. Fenichel, Asymptotic Stability with Rate Conditions. Indiana Univ. Math. J. 23 (1974), 1109 - 1137.

[17] N. Fenichel, Asymptotic Stability with Rate Conditions II. Indiana Univ. Math. J. 26 (1977), 81 - 93.

[18] N. Fenichel, Geometric Singular Perturbation Theory for Ordinary Differential Equations. J. Diff. Eqns. 31 (1979), 53 - 98.

[19] K. P. Hadeler, Selektionsmodelle in der Populationsgenetik, in "Methoden und Verfahren der Mathematischen Physik, Band 9" (eds. B. Brosowski, E. Martensen), 136 - 160, Bibliographisches Institut, Mannheim 1973.

[20] K. P. Hadeler, Mathematik für Biologen. Springer, Berlin, Heidelberg, New York 1974.

[21] J. K. Hale, Ordinary Differential Equations. Wiley - Interscience, New York, London, Sydney, Toronto 1969.

[22] J. K. Hale and P. Massatt, Asymptotic Behavior of Gradient - Like Systems, in "Dynamical Systems II" (eds. A. R. Bednarek, L. Cesari), 85 - 101, Academic Press, New York, London 1982.

[23] C. J. Harris and J. F. Miles, Stability of Linear Systems: Some Aspects of Kinematic Similarity. Academic Press, New York, London 1980.

[24] P. Hartman, Ordinary Differential Equations. Wiley & Sons, New York, London, Sydney 1964.

[25] U. an der Heiden, On Manifolds of Equilibria in the Selection Model for Multiple Alleles. J. Math. Biol. 1 (1975), 321 - 330.

[26] D. Henry, Geometric Theory of Semilinear Parabolic Equations. Lecture Notes in Mathematics 840, Springer, Berlin, Heidelberg, New York 1981.

[27] M. W. Hirsch, C. C. Pugh and M. Shub, Invariant Manifolds. Lecture Notes in Mathematics 583, Springer, Berlin, Heidelberg, New York 1977.

[28] U. Kirchgraber and E. Stiefel, Methoden der analytischen Störungsrechnung und ihre Anwendungen. Teubner, Stuttgart 1978.

[29] U. Kirchgraber, The Geometry in the Neighborhood of an Invariant Manifold and Topological Conjugacy. Research Report 83 - 01, Sem. Angew. Math., ETH Zürich 1983.

[30] H. W. Knobloch and F. Kappel, Gewöhnliche Differentialgleichungen. Teubner, Stuttgart 1974.

[31] H. W. Knobloch and B. Aulbach, The Role of Center Manifolds in Ordinary Differential Equations, in "Equadiff 5" (ed. M. Greguš), 179 - 189, Teubner, Leipzig 1982.

[32] J. P. LaSalle, The Stability of Dynamical Systems. Soc. Ind. Appl. Math., Philadelphia 1976.

[33] J. P. LaSalle, Stability Theory for Difference Equations, in "Studies in Ordinary Differential Equations" (ed. J. K. Hale), 1 - 31, Math. Assoc. Amer. Studies in Math. Vol. 14, 1977.

[34] A. M. Ljapunov, Problème général de la stabilité du mouvement. Ann. Math. Studies Vol. 17, Princeton 1949 (Ann. Fac. Sci. Toulouse Vol. 9 (1907), 203 - 475; Soc. Math. Kharkov 1892).

[35] I. G. Malkin, Theorie der Stabilität einer Bewegung. Oldenbourg, München 1959 (Gos. Izdat. Tehn.-Teor. Lit., Moscow, Leningrad 1952).

[36] K. J. Palmer, Linearization near an Integral Manifold. J. Math. Anal. Appl. 51 (1975), 243 - 255.

[37] K. J. Palmer, Qualitative Behavior of a System of ODE near an Equilibrium Point - A Generalization of the Hartman - Grobman Theorem. Preprint 372, Inst. Angew. Math., Univ. Bonn 1980.

[38] W. T. Reid, A Prüfer Transformation for Differential Systems. Pacific J. Math. 8 (1958), 575 - 584.

[39] J. Stoer and R. Bulirsch, Introduction to Numerical Analysis. Springer, Berlin, Heidelberg, New York 1980.

Index